GLOBSTERS

MICHAEL NEWTON

Typeset by Jonathan Downes,
Cover and Layout by SPiderKaT for CFZ Communications
Using Microsoft Word 2000, Microsoft Publisher 2000, Adobe Photoshop CS.

First published in Great Britain by CFZ Press

**CFZ Press
Myrtle Cottage
Woolsery
Bideford
North Devon
EX39 5QR**

© CFZ MMXII

ISBN: 978-1-905723-40-9

For Robert Henry Rines
(1922-2009)

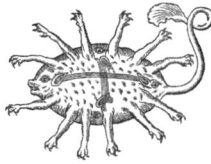

Contents

Foreword

aturalist, adventurer and Fortean author Ivan Terence Sanderson coined the term "globster" in 1962, to describe strange masses of organic tissue washed ashore by ocean tides. While Sanderson initially applied the term to one specific carcass, beached in western Tasmania two years earlier, today we know such strandings have occurred worldwide, with records spanning fifteen centuries. Nor is an ocean view required to spot a globster: certain lakes, as well, have vomited peculiar carcasses.

Globsters is the first attempt to survey all known "monster" strandings in a single dedicated volume, covering the years from 661 C.E. through 2010. In addition to 132 discoveries of lifeless remains, the book also examines thirty-six cases in which aquatic cryptids - "hidden animals," the subject of cryptozoology - were allegedly killed or captured alive by intrepid seamen or hunters. Some of the cases are globally famous, while others are virtually forgotten, known only from passing mention in documents covering other subjects. Certain globsters have been scientifically identified through DNA or other forms of testing; others are presumed to have prosaic explanations, although evidence is lacking; and a few remain profoundly enigmatic.

I owe thanks to Dr. Karl Shuker and Loren Coleman for advice received during my research for *Globsters*; to Ellen Hollis at the Bermuda National Library; to Henry Scannell, curator of Reference and Information Services at the Boston Public Library; to Oregon journalist James D. Adams; to L. Thurber, librarian at the Nantucket Atheneum; and to Jonathan Downes at CFZ Press for seeing the project to print. Any readers having information on the cases covered in this book - or others that I may have missed - are welcome to contact me through my website at www.michaelnewton.homestead. com.

1.
Something on the Beach

The seas and lakes give up their dead - but what, exactly, are they offering for our inspection? Across the ages, thousands of putrescent carcasses have fouled coastlines from East to West and pole to pole. For centuries before their flippant christening as "globsters," they inspired dread or religious awe and sometimes served as teaching tools for primitive marine biologists. So they remain today, although technology contrives to make them less mysterious.

The pitfalls of assessing globsters should be readily apparent. First, by definition, most are only *part* of something larger. On arrival, they are fragmentary, decomposed, and often partially devoured by who-knows-what. Those cast ashore at a remote location are unlikely to be viewed by trained observers - and, indeed, we have no means of estimating how many were never seen by human eyes at all. In cases where a carcass is discovered, unprofessional attempts at preservation may defeat the best intentions of an expert analyst arriving later on the scene. Worse yet, said "expert" analyst may judge a specimen without observing it at all, allowing preconceived ideas to speak for science.

Bryan Dunning, writing on his weekly Skeptoid podcast in May 2009, observed that "All too often, the default skeptical position has been to brush these [globsters] off as misidentified whale parts or some other marine life that has decayed to the point of being unrecognizable." While maintaining that most stranded carcasses may be identified by modern DNA profiling or other methods, Dunning rightly notes that "to be a responsible skeptic, you can't simply ignore the small number of cases that don't fit that explanation. The fact is that a few of these globsters are *not* consistent with the usual suspects, like whale blubber or basking shark carcasses, and that's something skeptics should be aware of."[1]

Even those globsters that are finally identified - and which may disappoint some diehard sea-serpent believers - are educational in themselves. As shown in Chapter 2, our knowledge of the giant squid, still sadly incomplete, comes almost totally from stranded carcasses. Cetaceans - whales, dolphins and porpoises - are also frequent stranders who reveal their secrets most often in death.

Whale Tales

Whales - the biblical Leviathan - were known from stranded carcasses five centuries before mankind began to hunt them systematically. Aristotle, in his *Historia Animalium,* wrote: "It is not known for what reason they run themselves aground on dry land; at all events it is said that they do so at times, and for no obvious reason." Today, with an average 2,000 strandings per year and ten cetacean species known for mass-beachings, debate over the cause of the phenomenon continues.[2]

Some cetacean species, chief among them the beaked whales (family *Ziphiidae*) are known primarily - or solely - from carcasses stranded on shore. Examples include[*]:

- Andrews' beaked whale (*Mesoplodon bowdoini*), also known as the Bowdoin's beaked whale, splay-toothed beaked whale or deepcrest beaked whale. Named in 1908, this species is known from thirty-five strandings on Australia, New Zealand, and neighboring islands. No specimen has yet been seen alive.[3]

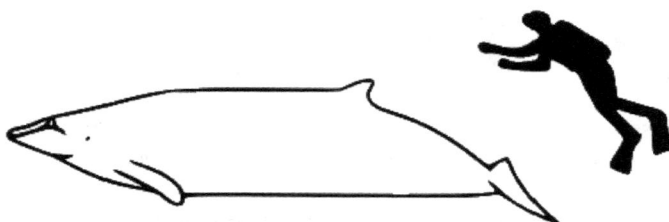

- True's beaked whale (*Mesoplodon mirus*), also known as the wonderful beaked whale. Named in 1913, it was believed to be exclusively a North Atlantic species until 1959, when a specimen was beached on the coast of South Africa. As with Andrews' beaked whale, no members of the species have been observed at sea.[4]

* These images, by the way, are by Chris Huh, courtesy of Wikimedia Commons

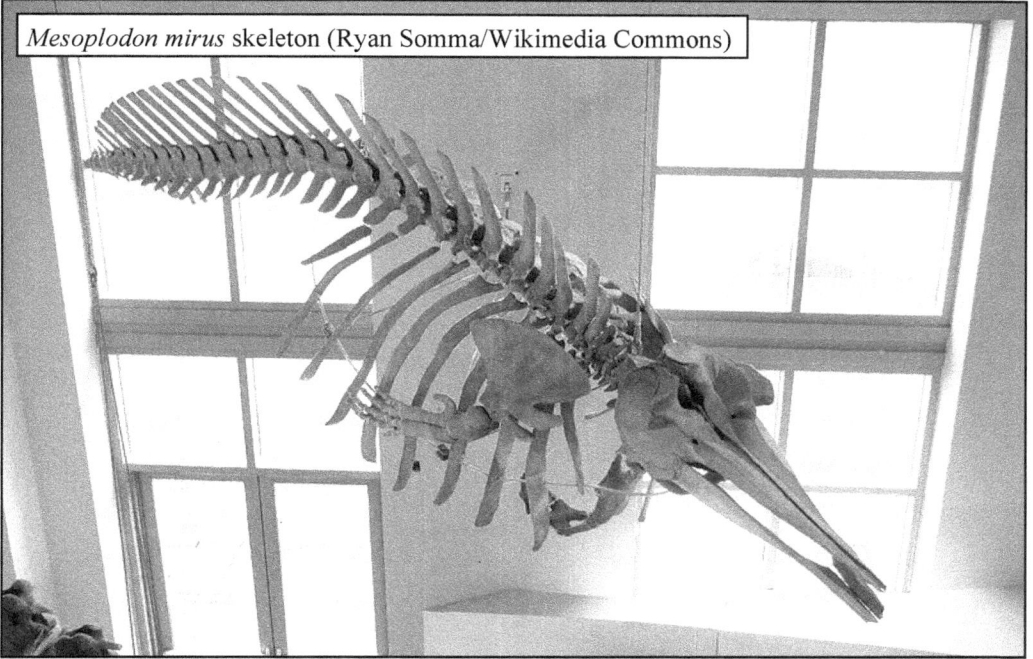

Mesoplodon mirus skeleton (Ryan Somma/Wikimedia Commons)

- Longman's beaked whale (*Indopacetus pacificus*), also known as the Indo-Pacific beaked whale and the tropical bottlenose whale. Named in 1926, from a skull beached on Queensland's northeastern coast in 1882, the species was not documented by another specimen until 1955, when a second skull was found at a fertilizer factory in Mogadishu, Somalia. Measurement of those skulls suggests a length of twenty-three feet for Longman's beaked whale, confirmed by live sightings from the Indian Ocean, the Pacific and the Gulf of Mexico since 1980.[5]

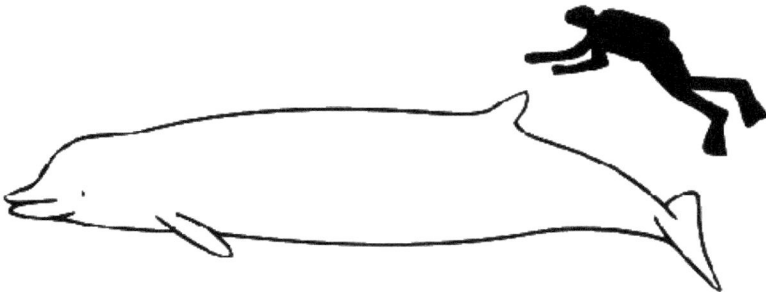

- Sowerby's beaked whale (*Mesoplodon bidens*), also known as the North Atlantic or North Sea beaked whale. The first species of beaked whale scientifically described, it was named in 1804 from a skull beached at Scotland's Moray Firth in 1800. While seen alive over the intervening centuries, Sowerby's beaked whale is still known primarily from 100-odd strandings.[6]

- Gervais' beaked whale (*Mesoplodon europaeus*), also known as the Gulf Stream beaked whale, Antillian beaked whale, and European beaked whale. Named in 1855 from a specimen stranded in 1840, this species was not seen alive until 1998. Only ten additional live sightings have been logged since then.[7]

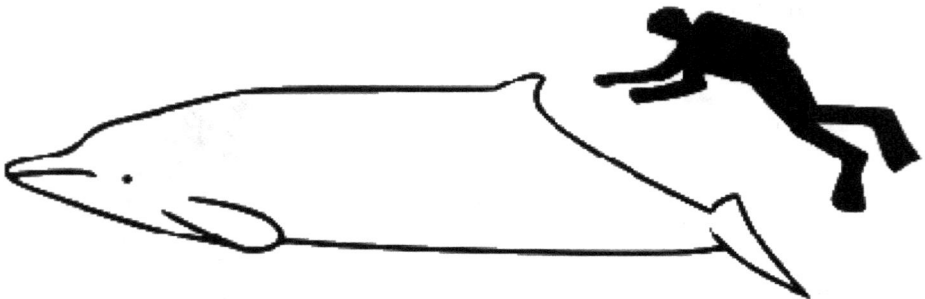

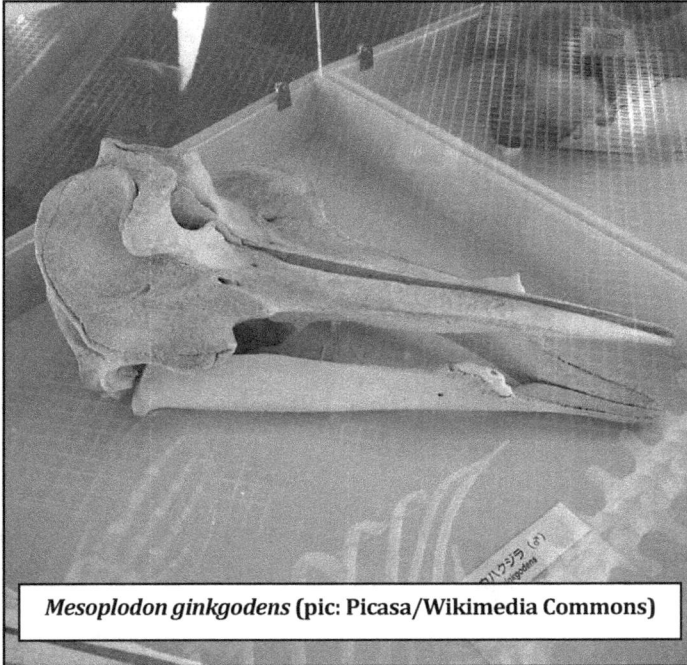

Mesoplodon ginkgodens (pic: Picasa/Wikimedia Commons)

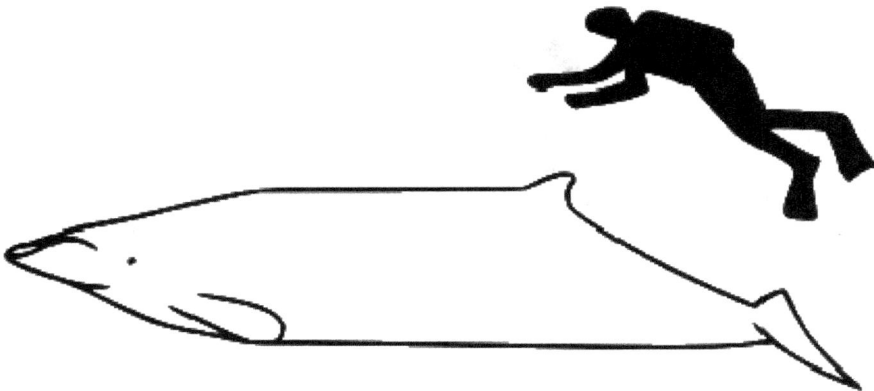

- The ginko-toothed beaked whale (*Mesoplodon ginkgodens*), also called the ginko beaked whale or Japanese beaked whale. Named in 1958, the species is known from fewer than two dozen widely-separated strandings and captures in the Pacific and Indian Oceans. No information exists on global abundance of the species.[8]

- The spade-toothed whale (*Mesoplodon traversii*), named in 1874 for a partial jaw found on the shore of Pitt Island in the Chathams two years earlier. Thereafter, it was lumped together with the strap-toothed whale (see below) until DNA testing linked the partial jaw to skullcaps found on White Island in the 1950s and on Chile's Robinson Crusoe Island in 1986. No living specimen has been observed.[9]

The world's first intact specimens were found on a beach in New Zealand as this book was going to press. For the first time, thanks to this and other photographs issued by the New Zealand Government, we know what the creature actually looks like. Details of its biology, however, are still completely unknown.

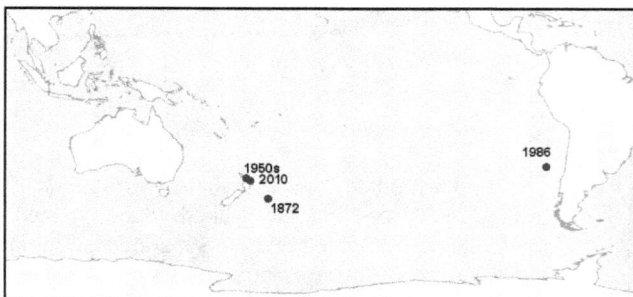

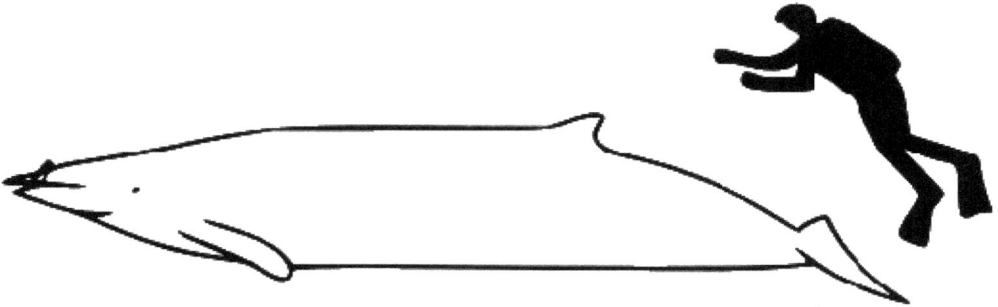

- The strap-toothed whale (*Mesoplodon layardii*), also known as Layard's beaked whale or the long-toothed whale. Named in 1865 from a skull beached on South Africa's coast, this twenty-foot squid-eater is known from frequent strandings and rare live sightings in the Southern Hemisphere. Its global population is unknown, but presumed to be substantial.[10]

- Hector's beaked whale (*Mesoplodon hectori*), another denizen of the Southern Hemisphere. Named in 1871, the species is known entirely from beached specimens. Two "probable" sightings occurred off California, in 1976 and 1978, but no live specimen has been observed since then.[11]

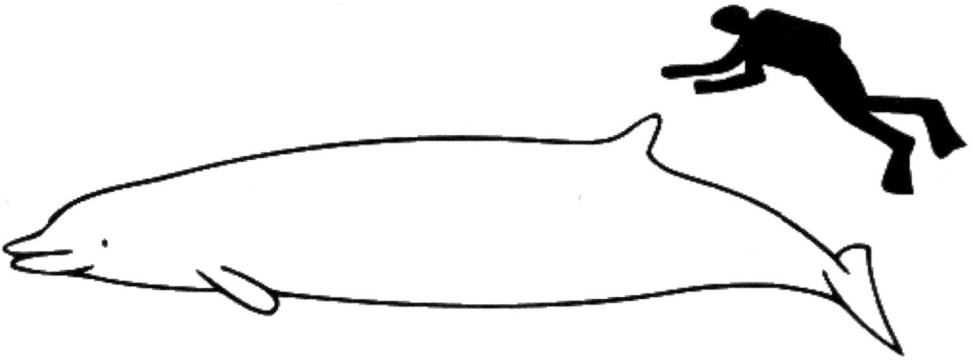

- Shepherd's beaked whale (*Tasmacetus shepherdi*), also known as the Tasman whale or Tasman beaked whale. Named in 1937 from its first known stranded specimen, beached on New Zealand's coast in 1933, the species yielded an additional twenty-seven carcasses by 2004. A handful of possible live sightings remain unconfirmed.[12]

- Stejneger's beaked whale (*Mesoplodon stejnegeri*), also called the Bering Sea beaked whale or saber-toothed whale. Described from a skull in 1885, nothing more was known of the species until the late 1970s, when a series of strandings occurred in the Sea of Japan and on California's coast. Even then, the whale's external appearance was not described until 1994, based on fresh strandings.[13]

  This image is from Wikimedia Commons and was contributed by user: OpenCage.

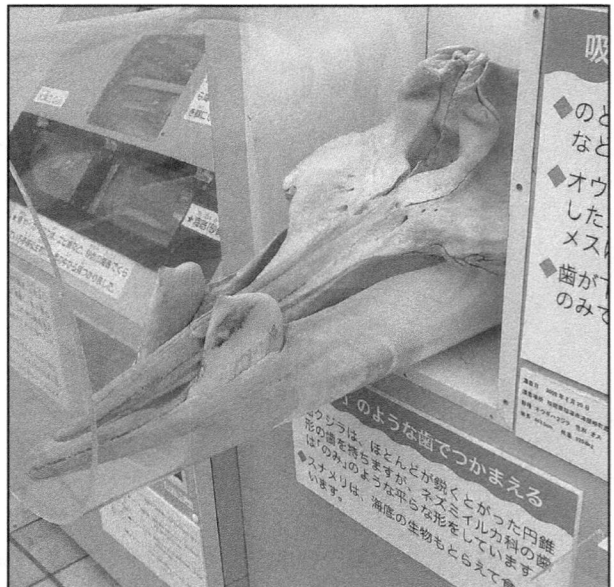

- Cuvier's beaked whale (*Ziphius cavirostris*), recognized as the world's most widely distributed beaked whale. Regarded as a monster in the Middle Ages, the species was described by Baron Georges Cuvier in 1823, from a French skull found in 1804. Its global range is known chiefly from strandings.[14]

- The pygmy beaked whale (*Mesoplodon peruvianus*), also known as the bandolero beaked whale, Peruvian beaked whale and lesser beak whale. The smallest of the Mesoplodonts, this specimen was first described in 1991, from a carcass beached on the coast of Baja California a year earlier. Assumptions that it only occupied the Western Hemisphere were shattered in October 1993, when a specimen washed ashore near Kaikoura, New Zealand.[15]

ABOVE: A whale stranded at Katwijk, Holland, in 1598.

BELOW: A stranded basking shark, as depicted in *Harper's Magazine,* in 1868.

Land Sharks

Cetaceans are not the only oceanic creatures prone to stranding, as we shall discover. Sharks are cast ashore quite frequently - witness the stranding of five basking sharks on Cornwall's coast within a single week of June 2004 - though it appears that no one bothers to compile global statistics.[16]

Stranded sharks are relevant to globsters because their skeletons - like those of the related skates and rays - consist entirely of cartilage and connective tissue. When a dead shark decomposes, the carcass may in some cases assume a form unrecognizable to amateur observers, possibly mistaken for a long necked creature resembling a plesiosaur. The placoid scales or dermal denticles of sharkskin may, in some cases, resemble bristling hair.

With that in mind, six large species of shark may present convincing globsters if beached in a state of advanced decomposition. They include, in order of descending size:

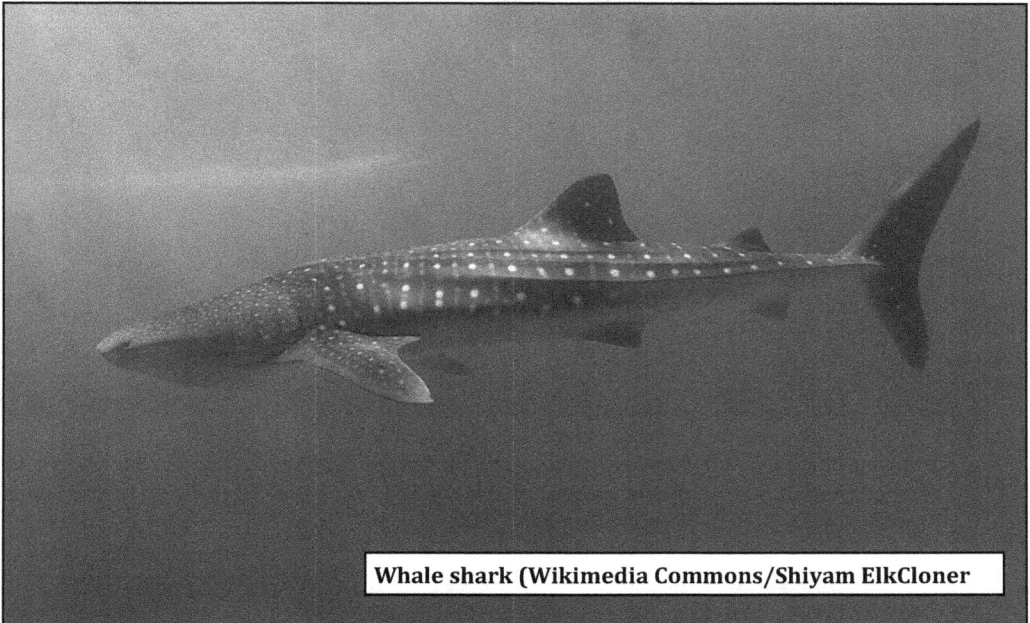

Whale shark (Wikimedia Commons/Shiyam ElkCloner

- Whale sharks (*Rhincodon typus*), the largest known living fish, with the record specimen measured at 41.5 feet long and weighing 47,000 pounds, caught off the coast of Pakistan in 1947. Unconfirmed reports of larger specimens abound. Irish naturalist Edward Perceval Wright claimed personal sightings of whale sharks exceeding forty-nine feet off the Seychelles, in 1868, and cited native tales of specimens approaching seventy feet. American ichthyologist Hugh McCormick Smith, writing in 1925, described a whale shark caught in Thailand that was fifty-six feet long and weighted 82,000 pounds. In 1934, crewmen aboard the *Maurguani* claimed their ship had struck a whale shark fifty-one feet long in the South Pacific.[17] Any one of those giants, suitably reduced by decomposition and scavengers, would make an impressive globster.

Basking shark (Chris Gotschalk /Wikimedia Commons)

- Basking sharks (*Cetorhinus maximus*) are the second largest living sharks known to science, found in temperate oceans worldwide. Like whale sharks, they are filter feeders, completely harmless to humans.

The largest accurately measured specimen, caught in Canada's Bay of Fundy in 1851, measured forty feet four inches and weighed roughly seventeen tons. Reports from Norway describe three specimens exceeding thirty-nine feet, with the largest forty-five feet long, but most adults caught in modern times range from twenty to thirty feet long.[18]

Proposed more often than its larger cousin as a globster candidate, *Cetorhinus maximus* may indeed be responsible for many recorded cases - but others, despite insistence from debunkers, would require the existence of basking sharks far beyond forty feet long.

Great white shark (Sharkdriver68/Wikimedia Commons)

• Great white sharks (*Carcharodon carcharias*) are the stereotypical man-eaters of *Jaws* and countless other melodramas, a reputation borne out by the International Shark Attack File's tabulation of species involved in attacks on humans.[19] Adult specimens commonly range from thirteen to seventeen feet in length, with most authorities acknowledging that some exceed twenty feet.

That said, the claim of a thirty-six-foot specimen caught near Port Fairy, Victoria, Australia, was later proved to be an error. Examination of the shark's jaw suggested a length closer to seventeen feet, and claims of a thirty-seven-footer caught off New Brunswick, Canada, in the 1930s are likewise widely doubted.[20]

Experts disagree concerning the largest reliably measured white shark on record: J.E. Randall cites a 19.5-foot specimen caught off Western Australia in 1987; the Canadian Shark Research Center claims a twenty-foot female caught off Prince Edward Island in 1988; a twenty-one-foot specimen weighing 7,330 pounds was allegedly verified by marine biologists John McCosker and Timothy Tricas in 1984.[21]

Great hammerhead (Jim/Wikimedia Commons)

- Great hammerheads (*Sphyrna mokarran*), denizens of tropical and warm temperate waters worldwide, may rival great whites for length, with a record specimen measuring twenty feet, but they are typically more slender. The heaviest specimen on record, measured at fourteen feet three inches, tipped the scales at 1,280 pounds.[22]

- Megamouth sharks (*Megachasma pelagios*), a model of which is pictured here by, 'opencage' courtesy of Wikimedia Commons unknown to science before 15 November 1976, when a U.S. Navy ship snagged the first specimen on its anchor chain, twenty-five miles offshore from Oahu, Hawaii. The new shark, another harmless filter feeder, measured fourteen feet six inches and weighed 1,650 pounds. By June 2010, another thirty-six specimens had been caught, with eleven stranded on beaches worldwide, and two reportedly observed at sea. While relatively sparse, the final tally seems remarkable, considering

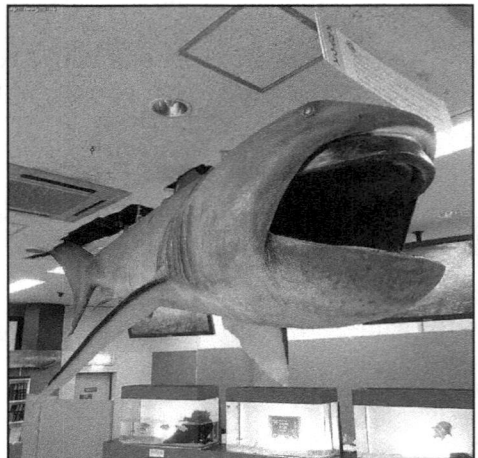

how long the megamouth was able to conceal itself from humankind. Indeed, Taiwanese fishermen caught three megamouths in May and June 2005 alone. One of those reportedly measured twenty-three feet long.[23]

- Tiger sharks (*Galeocerdo cuvier*), ranked second after great white sharks as man-eaters, prove elusive when it comes to record lengths. Adults routinely grow to fourteen feet, and all sources admit record lengths exceeding sixteen feet. Beyond that, reports of a twenty-four-foot female caught in 1957 and a twenty-two-footer hooked in 1922 remain officially unconfirmed.[24]

Tiger shark (Albert Kok/Wikimedia Commons)

Analyzing Globsters

Simple observation may not be enough to tag a globster with its proper name. Aside from shapeless heaps of flesh and rotting sharks that look like ancient reptiles, even carcasses of a familiar species may confound beachcombers, as demonstrated by a recent case from England.

On 20 September 2010, reporter Naomi Lloyd informed viewers of ITV's West Country bulletin that a polar bear had washed ashore near Bude, in Cornwall. "The bear comes from the Arctic Circle," Lloyd declared, "and an investigation is under way as to how it could have ended up there." How indeed, after a journey of some 2,700 miles? The mystery was solved when cooler heads recognized the "polar bear" as a dead cow, bleached white by immersion in salt water.[25]

If such a common animal can be mistaken for an alien invader, what may hasty or uneducated witnesses suspect when suddenly confronted with the rotting carcass of a creature rarely seen ashore, if ever seen at all? If veteran fishermen and sailors still report sightings of sea-serpents and other unidentified swimming objects, what are mere landlubbers to think of a decomposed whale, shark, or giant squid?

The risk of mistaken identification is clear, but another risk lies in assuming that all beached remains must be those of some beast known to science. When the world's fourth- or fifth-largest shark can hide from human eyes for centuries on end, what else is lurking in the sea? A sampling of media items from recent years suggests the scope of new discoveries in waiting.

- 18 September 2006: Spokesmen for Conservation International announced the discovery of fifty new marine species - including two new sharks that walk on their fins - from the waters off northwestern Papua New Guinea. Expedition leaders described the 45-million-acre region as "Earth's richest seascape" and "the most biodiverse marine area on the planet."[26]
- 5 February 2007: Members of the Panglao Marine Biodiversity Project announced results of a two-year survey conducted around the Philippine island of Panglao, located 390 miles southeast of Manila. They declared that "Numerous species were observed and photographed alive, many for the first time, and it is estimated that between 150 and 250 of the crustaceans and 1,500 to 2,500 of the molluscs are new species."[27]
- 8 March 2007: Marine biologists from the Smithsonian Institution reported their discovery of "a biodiversity bounty" on Panama's Pacific coastline. Without citing specific figures, members of the eleven-day expedition reported that "approximately 50 percent of the organisms found in some groups are new to science."[28]
- 8 October 2008: Australia's Commonwealth Scientific and Industrial Research Organization announced discovery of 274 new marine species living two kilometers beneath the surface of the Indian Ocean, at a site 130 miles south of Tasmania.[29]
- 4 October 2010: Spokesmen for the First Census of Marine Life announced the results of a decade-long, $669 million global survey that produced 30 million observations of 120,000 species at sea. More than 6,000 "potentially new species" were recorded, with 1,200 described in detail. Even then, the scientists estimated that another 750,000 marine species remain undiscovered.[30]

None of those finds was a giant sea-serpent, but the continuing march of new discoveries at sea renders invalid any skeptic's claim that the next beached globster "must be" a dead shark or a heap of whale blubber. The arrogant assertion that aquatic "monsters" don't exist because they *can't* exist is clearly false.

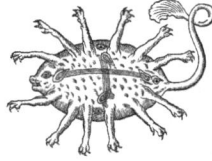

2.
Kraken

I t may be the most ferocious sea beast ever conjured from the mind of man, albeit with a helping hand from nature. A thing with many arms, so huge that it might be mistaken for an island until sailors trespassed on its flesh and were consumed, together with their vessels. Even when retreating from contact with humans it was dangerous, creating a vast whirlpool capable of sucking ships into the depths.

Its name was *Kraken*.

Swedish naturalist Carl Linnaeus may have been the first to use that name in print, in his *Systema Naturae* (1735), but similar beasts known as *hafgufa* and *lyngbakr* appeared in the Norwegian text *Konungs skuggsjá* ("king's mirror") five centuries earlier. Linnaeus ranked the Kraken among cephalopods and gave it the scientific name *Microcosmus*. Two decades later, Bishop Erik Pontoppidan's *Natural History of Norway* (1752-53) described the Kraken as a beast "the size of a floating island." Pontoppidan added, "It is said that if it grabbed the largest warship, it could manage to pull it down to the bottom of the ocean." According to Pontoppidan, a young Kraken once washed ashore at Alstahaug, in Nordland. Despite its fearsome reputation, Pontoppidan averred that Norwegian fishermen sought Krakens out, because fishing above the submerged monsters promised a bountiful catch.[1]

Swedish sailor, clergyman and author Jacob Wallenberg (1746-78) offered a typical description of the Kraken in his book *Min son på galejan* ("My son on the galley"). Therein, he wrote:

> Kraken, also called the Crab-fish, which is not that huge, for heads and tails counted, he is no larger than our Öland [Sweden's second-largest island] is wide [9.9 miles] ... He stays at the sea floor, constantly surrounded by innumerable small fishes, who serve as his food and are fed by him in return: for his meal, if I remember correctly what E. Pontoppidan writes, lasts no longer than three

months, and another three are then needed to digest it. His excrements nurture in the following an army of lesser fish, and for this reason, fishermen plumb after his resting place ... Gradually, Kraken ascends to the surface, and when he is at ten to twelve fathoms, the boats had better move out of his vicinity, as he will shortly thereafter burst up, like a floating island, spurting water from his dreadful nostrils and making ring waves around him, which can reach many miles. Could one doubt that this is the Leviathan of Job?[2]

Various theories were advanced to explain tales of the Kraken. Some texts named whales as the probable suspect, while others suggested undersea volcanoes, rogue currents, and newly-surfaced islets. French naturalist Pierre Denys de Montfort came closer to the truth in his *Histoire Naturelle Générale et Particulière des Mollusques* (1801-02), with a description of two hypothetical giant mollusks. One he called the kraken octopus; the other was a *Poulpe Colossal,* or colossal octopus, which he credited with sinking ships at sea. His list of incidents included an attack on a merchant vessel off the coast of Angola, and ten ships lost without a trace in the Caribbean, in April 1782.[3]

While the British admiralty offered a more prosaic explanation for the latter incident, Montfort supported his claims with descriptions of two large "octopus" tentacles described by members of the Dunkirk whaling fleet. One fragment, vomited up by a dying sperm whale, measured thirty-five feet long and bore suckers "broader than a hat." The other, initially mistaken for a sea-serpent, was even larger: forty-five feet long, and thirty inches in diameter.[4]

Sadly, Montfort did not live to see his supposition partly validated by discovery of a real-life Kraken. Disgraced and ridiculed, he died in Paris, penniless, in 1820. Thirty-five more years elapsed before science accepted the existence of a true giant cephalopod.

Enter *Architeuthis*

In retrospect, the monster should have come as no surprise to anyone. Globsters had paved the way, the first known carcass beached near Malmö, Denmark (later Sweden), in 1545. Locals tagged the nine-foot carcass as a *soemonk* - a monstrous fish resembling a robed Catholic monk.[5]

Iceland hosted the next carcass, stranded at Thingøresand in autumn 1869. It sported seven "tails," each roughly seven feet long, while an eighth approached thirty feet. A third specimen, beached at Ireland's Dingle Bay in October 1673, measured nineteen feet overall. Yet another washed ashore at Ulvangen Fjord, Norway, in 1680, but no measurements were recorded.[6]

The eighteenth century produced six more Krakens: an unmeasured specimen at Jutland, Denmark, in 1770; a carcass with a seven-foot body, found floating off Newfoundland's

ABOVE: Pierre Denys de Montfort's *Poulpe Colossal*.

BELOW: An early photo of a stranded *Architeuthis.*

Grand Banks in May 1785; a monster close to forty feet long, beached at Arnarnaesvik, Iceland, in November or December 1790; two more Newfoundland specimens found sometime in the early 1790s; and another Danish stranding in 1798.[7]

Discoveries accelerated after 1800, with five finds in the first half of the nineteenth century. In 1802 a cephalopod six to seven feet long was found off the coast of Tasmania. Another, weighing 400 pounds, was hauled from the Atlantic in 1817. A set of beaklike jaws alone, weighing some 176 to 187 pounds, came ashore at Raabjerg, Denmark, in December 1853. The final weeks of 1855 offered two specimens: a partial arm and gladius (a horn-like internal body part) measuring twelve feet overall, found floating off the Bahamas on 5 November; and an unmeasured carcass beached at Aalbækstrand, Denmark, in December.[8]

Finally, science was prepared to acknowledge the Kraken's existence. More specifically, Danish zoologist Japetus Steenstrup was prepared, in 1857, to identify the beast as *Architeuthis dux* ("Atlantic giant squid"). Somewhat ironically, the shift from disbelief to scientific acceptance prompted a veritable frenzy of naming new giant squid species. Steenstrup himself added a second species, *A. monachus,* to accommodate the carcass found at Aalbækstrand in 1855.[9] By 1912, no less than seventeen additional species had been hypothetically identified worldwide. Listed alphabetically, they include:

- *Architeuthis bouyeri,* named in 1862 for a squid found near Tenerife, in the Canary Islands, in the same year;
- *A. clarkeii,* named in 1933 for a specimen found on 14 January of that year, at Scarborough Beach in Yorkshire, England;
- *A. hartingii,* named in 1860 for a specimen of unknown origin, found the same year;
- *A. harveyi,* named in 1874 for a nineteen-foot specimen found in October 1873, at Newfoundland's Conception Bay;
- *A. japonica,* named in 1912 for a squid found that year in Tokyo Bay;
- *A. kirkii,* named in 1887 for a beak and club from one tentacle, beached at Campbell Bay, New Zealand, on 30 June 1886;
- *A. longimanus,* named in 1888 for a beak and its surrounding buccal mass found at New Zealand's Lyall Bay in October 1887;
- *A. martensi,* the "North Pacific giant squid," named in 1880 for partial remains measuring 12.5 feet, purchased at a fish market in Yedo, Japan, in 1873;
- *A. megaptera,* named in 1878 for specimen beached at Cape Sable, Nova Scotia, sometime in 1870;
- *A. mouchezi,* named in 1875, to describe a partial specimen stranded on St. Paul Island, in the Indian Ocean, on 2 November 1874;
- *A. nawaji,* named in 1935 for remains snagged by a trawler at Île d'Yeu, off the Vendée coast of western France, on 26 June of that year;
- *A. physeteris,* named in 1900 from several jaws and a mantle vomited by a sperm whale near Terceira Island, in the Azores, on 18 July 1895;
- *A. princeps,* named in 1875 for a fifty-two-foot giant reported from Newfoundland's
- Strait of Belle Isle;

- *A. proboscideus,* named in 1875 to describe the Dingle Bay squid beached in 1673;
- *A. sanctipauli,* the "Southern giant squid," another species named in 1875, for the St. Paul Island specimen of November 1874;
- *A. stockii,* named in 1882 for a mutilated carcass beached at Lyall Bay, New Zealand, on 23 May 1879;
- *A. verrilli,* named in 1882 for a twenty-five-foot squid stranded at Island Bay, New Zealand, on 6 June 1880.[9]

That plethora of *Architeuthis* species is hotly disputed today. In 1984, Smithsonian Institution zoologist Clyde Roper wrote that "the 19 nominal species can in fact be encompassed by only three: *Architeuthis sanctipauli* in the Southern Hemisphere, *A. japonica* in the northern Pacific, and *A. dux* in the northern Atlantic." Seven years later, marine biologist Frederick Aldrich, from Memorial University of Newfoundland, wrote: "I reject the concept of 20 separate species, and until that issue is resolved, I choose to place them all in synonymy with *Architeuthis dux* Steenstrup."[10] Martina Roeleveld of the South African Museum agreed with Aldrich in 1996, while Australian marine biologist Mark Norman sided with Roper in 2000, writing: "The number of species of giant squid is not known although the general consensus amongst researchers is that there are at least three species, one in the Atlantic Ocean (*Architeuthis dux*), one in the Southern Ocean (*A. sanctipauli*) and at least one in the northern Pacific Ocean (*A. martensi*)."[11]

One species or three, *Architeuthis* was known from 592 confirmed specimens by the end of 2006. Those included 306 from the Atlantic, 264 from the Pacific, twenty from the Indian Ocean, and two from the Mediterranean Sea.[12] Others have been seen or caught over the years, but the first adult giant squid was not photographed alive at sea until 30 September 2004, when researchers from the National Science Museum of Japan and the Ogasawara Whale Watching Association snapped 556 photos at a depth of three thousand feet, off Japan's Ogasawara Islands. Two years later, near the same location, members of the same team made the first videotape of a live adult *Architeuthis.* They snared that specimen, but it died while being reeled in.[13]

Even now - 277 years since Carl Linnaeus tried to classify the Kraken, and 155 years since Japetus Steenstrup graced it with a Latin name - our knowledge of the giant squid is sparse. One of the most intriguing questions, at least for cryptozoologists, is the unsettled issue of maximum size.

How Big is "Giant"?

Squids are commonly measured in sections, particularly when decomposed or mutilated remains arrive on shore as globsters. Mantle length (ML) measures the squid's body from its tail to the back of its head. Head length (HL) covers the relatively small distance from the mantle to the base of the arms and tentacles. Arm length (AL) measures a squid's eight arms from base to tip. Tentacle length (TL) measures the two feeding tentacles, including their terminal "clubs" if intact. In some species of squid, tentacle length exceeds the other measurements combined. When a complete specimen is found or captured, its whole length (WL) measures the animal from its tail to the tip of its extended tentacles. Alternatively, standard length (SL) measures from tail to the tips of the arms, excluding the longer tentacles.

But how big do the deep-sea giants grow?

Writing in 2008, authors Steve O'Shea and Kat Bolstad note that "*Architeuthis* is frequently misreported to attain a total length of 20 metres (~65 feet)," then proceed to debunk that claim by means of what some readers may regard as an unwarranted assumption. Referring to a New Zealand female specimen measured at fifty-five feet two inches in 1887, and described by its contemporary observers as "in all ways smaller than any of the hitherto-described New Zealand species," O'Shea and Bolstad declare that "this simply cannot be correct, and this length almost certainly is a product of imagination or lengthening (stretching like rubber bands) of the very slight tentacular arms, as it mantle was only 71 inches long (1.8m)." The true WL measurement of "a comparable-sized female," they maintain, should be around thirty-two feet.[14]

Even if we concede the unfounded assumption of "stretching" for one specimen in 1887, what does it prove about maximum length for live Architeuthids at sea? O'Shea and Bolstad say that they have personally examined 130 *Architeuthis* specimens, finding none with an ML greater than seven feet three inches or a TL exceeding forty-two feet.[15]

Again, conceding that to be the case, what does it prove?

Larger squids have been described throughout recorded history, and even disallowing eyewitness accounts of "ones that got away," remaining evidence suggests that some giant squids do surpass the hypothetical maximum imposed by O'Shea and Bolstad. Examples include:

- November/December 1790: Observers of the Arnarnaesvik specimen reported that its "body right from the head" measured twenty-one feet five inches.
- November 1861: The mantle of a specimen found off Tenerife was roughly measured at fifteen to eighteen feet, more than twice the O'Shea-Bolstad maximum.
- Winter 1870: A squid with a WL of forty-seven feet beached at Lamaline, Newfoundland.
- October 1871: Fishermen on Newfoundland's Grand Banks snared a squid with a fifteen-foot mantle.
- Autumn 1872: Richard Ellis notes a fifty-two-foot specimen found alive at Coomb's Cove, Newfoundland.
- December 1872: Ellis reports a squid forty-six feet long, washed ashore at Newfoundland's Bonavista Bay.
- October 1873: Ellis credits a specimen found off Newfoundland's Portugal Cove with a length of forty-four feet.
- 1875: Zoologist Addison Verrill described a fifty-two-foot *Architeuthis* found alive at Newfoundland's Strait of Belle Isle.
- November 1877: Verrill noted a forty-four-foot squid beached at Lance Cove, Newfoundland.

- November 1878: Verrill documented a fifty-five-footer hauled ashore at Thimble Tickle, Newfoundland.
- June 1880: Verrill cited another fifty-five-foot squid beached at New Zealand's Island Cove.
- October 1887: A squid stranded at Lyall Bay, New Zealand, measured fifty-seven feet.
- October 1924: Remains of a "champion octopus" - revealed by the remnants of ten arms to be a squid, in fact - washed ashore at Baven-on-Sea, near Margate, Natal, South Africa. Based on diameter of the ten stumps remaining, observers guessed that the arms would be fifty feet long. The globster as found was twenty-eight feet long, and its apparent head was nine feet in diameter.
- March 1928: Residents of Ranheim, Norway, found a forty-six-foot squid on shore.
- 1930: Residents of Goose Bay, New Zealand, measured a beached squid with an eleven-foot mantle and thirty-foot arms.
- 1989: Author Michael Bright, in *There Are Giants in the Sea*, writes that "[t]he largest authenticated giant squid found since 1900 was caught by the crew of a U.S. Coast Guard ship patrolling the Grand Bahamas Bank near Tongue of the Ocean....It measured 14.3m (47 feet)."[15]

Shall we reject those specimens on the assumption that they were measured incorrectly, or that the observers were lying? If we accept them, do those reports define the growth limits of *Architeuthis?*

Richard Ellis, writing in 1999, refers to an "accepted maximum known length of 55 feet" for giant squids. Authors Nettie and George MacGinitie, in their *Natural History of Marine Animals* (1949), report that "two arms of *Architeuthis* that were 42 feet long were found, and if one reconstructed a body...the squid to which these arms belonged was 4.6 feet in diameter and 24 feet long, with an overall measurement of 66 feet."[16]

Witness J.D. Starkey described his meeting with an even larger squid in a September 1963 article for *Animals* magazine. While standing night watch on a 175-foot trawler off the Maldive Islands, in the Indian Ocean, Starkey claimed to have seen a squid nearly as long as the ship.[17]

Following publication of his book on giant squids in 1998, Richard Ellis received a letter from one Dennis Braun, who claimed a sighting of another huge squid in early 1969, at Isla de Vieques, Puerto Rico. Observing it in clear water, from the deck of a U.S. Navy ship, Braun estimated its length as "probably at least 100 feet."[18]

Even fragmentary remains tell a story. Frank Bullen, in his book *The Cruise of the Cachalot: Round the World After Sperm Whales* (1898), describes a piece of tentacle vomited by the first whale his crew harpooned in 1875. He described it as:

> "...a massive fragment of cuttle-fish - tentacle or arm - as thick as a stout man's body, and with six or seven sucking-discs or acetabula on it. These were about as large as a saucer, and on their inner edge were thickly set

with hooks or claws all round the rim, sharp as needles, and almost the shape
and size of a tiger's."

Bernard Heuvelmans estimated that the tentacle's owner must have dwarfed the squid caught
at Thimble Tickle three years after Bullen's adventure.[19]

The use of sucker-width to calculate the overall length of a missing squid inspires continuing
debate. Based on calculations from known specimens, Dr. Heuvelmans proposed the
following formula:

"If the body, including the head, of an *Architeuthis* is N meters long, then
the diameter of its largest suckers is about N centimeters."

Inversely, by measuring a sucker - or the scar left by one on a whale - we should be able to
calculate the minimum total length of a particular squid. Heuvelmans noted that:

"I specify *minimum size*, since nothing proves that a particular sucker is the
largest found on the animal."[20]

From there, Heuvelmans proceeds to quote British zoologist Leonard Harrison Matthews,
from *The Sperm Whale* (1938). After examining eighty-one specimens killed by whalers,
Matthews wrote:

"Nearly all male sperm whales carry scars caused by the suckers and claws
of large squids, scars by suckers up to 10 cm [3.94 in.] in diameter being
common."

From that, Heuvelmans calculates that a common-sized *Architeuthis* should measure thirty-
two feet six inches, or forty-eight feet nine inches with its tentacles extended.[21]

What, then, are we to make of reports describing much larger scars on the hides of sperm
whales? Zoologist Ivan Sanderson, writing in *Follow the Whale* (1956), reports that:

"[t]he largest rings from the largest squids have a diameter of about four
inches, yet scars left by such suckers on the skin of captured sperm whales
have measured over eighteen inches in diameter"

which, by the Heuvelmans formula, would suggest a squid at least 150 feet long. German-
American science writer Willy Ley, in *The Lungfish, the Dodo, & the Unicorn* (1948), also
described "something looking like the mark of a sucking disk more than two feet in diameter,"
found on a whale's hide, but Heuvelmans dismissed that statement as a misprint of "two
inches."[22]

Two arguments oppose the use of sucker scars to calculate the length of missing monsters.
One suggests that scars expand as a sperm whale grows, but Heuvelmans found that notion
"highly improbable," noting that "the largest scars are usually just as clear and thus just as

recent-looking as the smaller ones. It is also unknown whether very young sperm whales are also similarly marked."[23] Or, we might add, if they could survive an adult giant squid's attack.

A more plausible theory suggests that four-inch sucker scars may have been inflicted by some squid species other than *Architeuthis*. Only squids have suckers with serrated edges capable of scarring whales, but the ratios of sucker-width to total length varies dramatically among squid species. Heuvelmans specifically cites the genus *Sthenoteuthis,* whose thirty-one-inch specimens may sport suckers nearly an inch in diameter. A "giant" species - the purpleback flying squid, *Sthenoteuthis oualaniensis* - boasts a record mantle length of twenty-six inches and is found only in the northern Indian Ocean, in the region of the Red Sea, Gulf of Aden and Arabian Sea.[24]

Geographical problems aside, a Sthenoteuthid capable of inflicting four-inch scars should be roughly five times the record length acknowledged by science, approaching eleven feet in length. As Heuvelmans suggested, it seems more logical to seek an Architeuthid twice the recognized "official" length than to inflate its smaller relative five hundred percent.[25]

Colossal Cousins

That said, we face another question: Is *Architeuthis* the largest squid on Earth? A disputed contender for the title is *Mesonychoteuthis hamiltoni,* variously known as the colossal squid, Antarctic squid, or giant cranch squid. The species was discovered in 1925, after a sperm whale harpooned off the South Shetland Islands disgorged two partial tentacles. One measured 46.3 inches long, and its club featured unique swivel-based hooks that can rotate in any direction.[26]

Three decades passed before dying whales vomited remains of two more colossal squids, sometime in 1956-57. Observers estimated that one specimen, taken off the South Orkney Islands, would have measured thirty-nine feet. The other, from the South Shetlands, was judged slightly smaller, around thirty-two feet six inches. A survey published in 1999 claimed that 188 colossal squids had been caught so far, and seven more discoveries were reported between March 2003 and March 2008, for a minimum total of 195.[27]

The largest whole specimen of *Mesonychoteuthis* on record was caught alive by the fishing vessel *San Aspiring* in February 2007, in Antarctica's Ross Sea. Its captors pegged the squid's length at thirty-three feet, with a weight of 992 pounds, but scientists who received its dead body at the Museum of New Zealand Te Papa Tongarewa adjusted those figures. Post-mortem freezing and thawing reduced the squid's overall length to fourteen feet, through shrinkage of

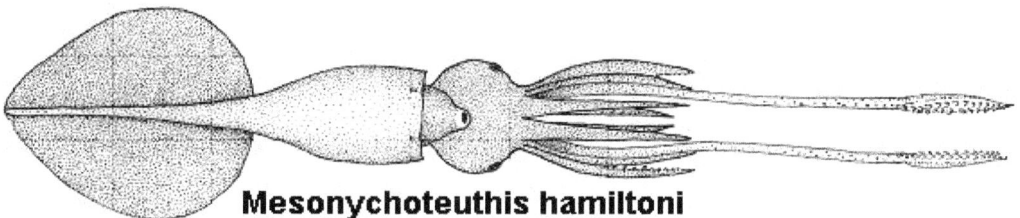

Mesonychoteuthis hamiltoni

tentacles, while it tipped the scales at 1,091 pounds. Researchers Steve O'Shea, Kat Bolstad and Tsunemi Kubodera determined that *Mesonychoteuthis* has a longer, heavier mantle than *Architeuthis* but sports shorter tentacles. Its eyes, with a diameter of twelve to sixteen inches in life, are the largest known in the animal kingdom. Despite those qualified findings, some published reports persist in calling the colossal squid Earth's "largest known invertebrate," with a hypothetical maximum length of forty-six feet.[28]

Other impressive squids, though not on the same scale as *Architeuthis* or *Mesonychoteuthis,* include:

- *Galiteuthis phyllura*, a species of glass squid related to the colossal squid (family *Cranchiidae*). In 1984, the Russian trawler *Novoulianovsk* brought up a record specimen from 3,900 feet in the Sea of Okhotsk. Dr. Kir Nesis examined the remains at Moscow's P.P. Shirshov Institute of Oceanology, Russian Academy of Sciences, pegging its total length at thirteen feet. In terms of mantle length alone, at 8.61 to 8.93 feet, he judged it the second-longest known squid.[29]
- *Moroteuthis robusta*, the robust clubhook squid, judged by some authors to be Earth's third-largest squid, after *Architeuthis* and *Mesonychoteuthis.* Its record mantle length is six feet six inches, while its arms may be equal in length to the mantle. A photograph of *M. robustus* snapped in Japanese waters and published in 1993 was initially believed to be the first-ever photo of an *Architeuthis* at sea.[30]
- *Megalocranchia fisheri*, another glass squid, with a record mantle length of 5.9 feet and a total length of 8.9 feet. Juveniles live near the surface, while adults inhabit mesopelagic depths (650 to 3,300 feet) by day and rise toward the surace at night.[31]
- *Dosidicus gigas*, variously known as the Humboldt squid, jumbo squid, or *diablo rojo* ("red devil," in Spanish). An aggressive predator inhabiting the Eastern Pacific's Humboldt Current, found from Tierra del Fuego to Alaska, *D. gigas* hunts in shoals that may exceed 1,200 individuals. While they normally thrive at mesopelagic depths, they may also be found near the surface. More than a thousand Humboldt squids beached themselves on Washington's Long Beach Peninsula in autumn 2004. Officially, the record mantle length for *D. gigas* is five feet eight inches, with a maximum overall length of ten feet.[32] That said, a recent investigation has raised some startling - even frightening - questions about the Humboldt squid.

In 2006, the History Channel's *MonsterQuest* program organized an expedition to the Sea of Cortez (or Gulf of California), located between Baja California and the Mexican mainland. The team included Dr. Roger Hanlon, senior scientist at the Marine Biological Laboratory in Woods Hole, Massachusetts; malacologist Scott Cassell, who survived an underwater attack by several Humboldt squids in 1996; and documentary producer Doug Hajicek. Their goal: to catch a Humboldt squid, affix a camera to its living body, and release it as bait in hopes of attracting an *Architeuthis.*

The results, aired initially on 14 November 2007, revealed a huge cephalopod rising from the depths to attack and consume the bait squid. Unfortunately, the camera angle did not permit identification of the predator. Analyst Peter Schmitz, at Indiana's Motion Engineering firm, used known proportions to determine that the larger squid was either a sixty-foot specimen of *D. gigas,* or a 108-foot specimen of *Architeuthis!* A follow-up broadcast on 8 October 2008 revealed no additional footage.[33]

- *Taningia danae*, the Dana octopus squid, named for the Danish research vessel *Dana,* which caught the first specimen in 1931. Dubbed an "octopus squid" (family *Octopoteuthidae*) because it lacks the two feeding tentacles found on other squid species, the record specimen of *T. danae* had a mantle 5.6 feet long and a total length of seen feet six inches.[34]
- *Kondakovia longimana*, commonly called the giant warty squid or longarm octopus squid, a hooked squid of the family *Onychoteuthidae.* First described in 1972, from three specimens caught near the South Orkney Islands, this squid apparently does not stray from the Southern Ocean. The largest specimen to date, measuring seven feet six inches overall, was found dead by members of the British Antarctic Survey in April 2000. Prior to that discovery, the maximum acknowledged length for *K. longimania* was three feet nine inches.[35]
- *Thysanoteuthis rhombus*, sole confirmed species of the genus *Thysanoteuthis,* known as the diamond squid or diamondback squid for the large triangular fins that extend the full length of the mantle. A denizen of tropical and subtropical seas, commercially fished in the Sea of Japan and around Okinawa, it commonly attains a mantle length of three feet three inches, with a record weight of sixty-six pounds.[36]

Nor are the mysteries surrounding squids at sea exhausted yet. In December 2001, Michael Vecchione from the National Museum of Natural History in Washington, D.C., announced discovery of a new and most peculiar squid inhabiting deep waters of the Gulf of Mexico, and the Pacific, Atlantic, and Indian Oceans. "We have never seen anything like it," Vecchione said. "It's a very weird-looking thing - really big fins, really long arms and this tiny little body in between." Instead of the characteristic eight arms and two feeding tentacles, the new squid sports ten indistinguishable appendages of identical length that radiate from the main axis of the body like spokes on a bicycle wheel, each with an elbow-like bend from which the remainder dangles.[37]

While no specimens of the mystery squid have yet been captured, eight have been observed from submersible vehicles worldwide since 1998. Size estimates vary widely, from five to twenty-three feet. Based on appearance alone, Vecchione tentatively assigned the species to the genus *Magnapinna* (big-finned squids), which presently includes three classified species. "They must be fairly common for people to bump into them all over the world," he said. "New species are a dime a dozen in the deep sea, and I suspect there are a lot of very weird things down there."[38]

If we learn nothing more from the Kraken's story, that advice should linger with us as we survey globsters on a global scale.

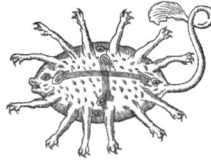

3.

Octopus Giganteus

I f Kraken-hunter Pierre Denys de Montfort mistook the giant squid for a "colossal octopus," does that mistake rule out the possible existence of a truly giant octopus species? Again, the evidence provided by a globster may suggest the opposite.

Modern science presently recognizes three species of "giant" octopus. The two most commonly labeled as giants - the North Pacific giant octopus and southern giant octopus - are

Enteroctopus dofleini

members of the genus *Enteroctopus*, with four species inhabiting temperate waters worldwide. The larger of the "giants," *Enteroctopus dofleini*, is found in coastal waters ranging from southern California around the North Pacific Rim to Japan, including the Bering Sea and Sea of Okhotsk. The second-largest species, *E. magnificus*, dwells along the southwestern coast of Africa, from Namibia to the vicinity of Port Elizabeth, South Africa.[1]

Of the two *Enteroctopus* "giant" species, *E. dolfeini* is arguably the largest recognized octopus on Earth. Reports of its maximum size vary widely, from an arm-span of fourteen feet and average adult weight of thirty pounds to claims of arms exceeding fifteen feet in length (for a spread topping thirty feet) and a record weight of 598 pounds.[2] Author Frank Lane, writing in 1960, claimed one specimen from Alaskan waters boasted an arm-span of thirty-two feet, but noted that "the body of this octopus was less than 18 in. long, and the arms were extremely tenuous toward their tips."[3] *E. magnificus* is significantly smaller, with arms reaching five feet in length, for a span exceeding ten feet.[4]

Another contender for "giant" status is the seven-arm octopus (*Haliphron atlanticus*), so-called because its eighth arm - the hectocotylus, specially modified for use in reproduction - is generally coiled in a sac beneath its right eye. The largest specimen confirmed to date, caught off the eastern coast of New Zealand in 2002, measured nine feet three inches long and weighed 134 pounds.[5]

Impressive species all, but for a true gigantic octopus we must turn back the calendar and join the residents of St. Johns County, Florida, as they confront a most surprising globster on a local beach.

Dr. Webb's What-is-It

Anastasia Island lies along the northeast Atlantic coast of Florida, its eighteen-mile length separated from the mainland by the Matanzas River - in fact, a saltwater estuary. Its lighthouse, once a watchtower for nervous Spanish settlers, dates from 1874. Part of the island lies within St. Augustine's city limits; its other modern settlements include St. Augustine Beach, Butler Beach, Crescent Beach, Anastasia and Coquina Gables.

On 30 November 1896 a strange visitor planted itself on St. Augustine Beach. Young bicyclists Herbert Coles (or Colee) and Dunham Coretter (or Coxetter) were first to spot the object, partially buried in sand. They thought it was a whale's carcass and rushed to notify Dr. DeWitt Webb, a local physician and amateur naturalist who had founded the St. Augustine Historical Society and Institute of Science. Webb and others viewed the carcass on 1 December, observing the apparent stumps of arms and thereupon deciding that the pinkish-white object was, in fact, a monstrous octopus. As described in the *New York Herald* on 2 December, "Its body, which is estimated to weigh about five tons, has sunk into the sand to a considerable depth, but that portion above the surface measures twenty-three feet in length, four feet in height, and more than eighteen feet across the widest part of its back."[6]

Inclement weather prevented further examination of the carcass until 5 December. Two days later, Webb returned with photographers Ernest Howatt and Edgar Van Horn, who snapped several exposures of the carcass. Another local, John L. Wilson, undertook private excavation around the remains, reporting exposure of tentacle fragments. As he told Webb, "One arm was lying west of the body, twenty-three feet long; one stump of an arm, west of the body, about four feet; three arms lying south of the body and from appearance attached to same (although I did not dig quite to body, as it lay well down in the sand, and I was very tired), longest measured over thirty-two feet, the other arms were three to five feet shorter."[7]

Sadly, Wilson worked alone, but his description was corroborated by South Beach hotelier George Grant, who penned a letter to his hometown newspaper, the *Pennsylvania Grit,* in Williamsport. Published on 13 December, it read:

> The head is as large as an ordinary flour barrel, and has the shape of a sea lion head. The neck, if the creature may be said to have a neck, is of the same diameter as the body. The mouth is on the under side of the head and is protected by two tentacle tubes about eight inches in diameter and about 30 feet long. These tubes resemble an elephant's trunk and obviously were used to clutch in a sucker like fashion any object within their reach. Another tube or tentacle of the same dimensions stands out on the top of the head. Two others, one on each side, protrude from beyond the monster's neck, and extend fully 15 feet along the body and beyond the tail. The tail, which is separated and jagged with cutting points for several feet, is flanked with two more tentacles of the same dimensions as the others and 30 feet long. The eyes are under the back of the mouth instead of over it. This specimen is so badly cut up by sharks and sawfish that only the stumps of the tentacles remain, but pieces of them were found strewn for some distance on the beach, showing that the animal had a fierce battle with its foes before it was disabled and beached by the surf.[8]

Dr. Webb, meanwhile, had sent his photos of the globster to Joel Asaph Allen, at Harvard University's Museum of Comparative Zoology. Allen never responded, but Webb's letter and photographs somehow found their way to Yale professor Addison Emery Verrill, widely regarded as America's top malacologist.[9] Impressed, Verrill wrote a hasty article for the January 1897 issue of the *American Journal of Science,* which read in part: "The proportions indicate that this might have been a squid-like form, and not an Octopus. The 'breadth' is evidently that of the softened and collapsed body, and would represent an actual maximum diameter in life of at least 7 feet and a probable weight of 4 or 5 tons for the body and head. These dimensions are decidedly larger than those of any of the well-authenticated Newfoundland specimens. It is perhaps a species of Architeuthis."[10]

The ink was barely dry on that article before Verrill changed his mind, in an article written for the *New York Herald'*s Sunday supplement on 3 January 1897. There, Verrill declared, "Dr. Webb has sent me photographs, four different pictures of the animal. They were taken on the same day he examined it. They show that the body is flattened, pear shaped, largest near the back end, which is broadly rounded and without fins. This form of the body and its proportions show that it is an eight-arm cuttlefish, or octopus, and not a ten-armed squid like the devil fish of other regions. No such gigantic octopus has been heretofore discovered."[11]

A storm swept the carcass back out to sea on 9 January, but it reappeared at Crescent Beach, two miles south of its original location, on 15 January. Dr. Webb's first effort to secure it failed, a gang of workmen unable to shift the carcass bare-handed, but a team of horses finally succeeded in dragging the hulk farther inland, where it was photographed once more.[12] A local tabloid newspaper, *The Tatler,* told its readers: "So far as can be determined at present, it belongs to no family not extinct, and is principally interesting on account of its great size, being about twenty-one feet long, without a head. Professor W.H. Dall of the Smithsonian Institute [*sic*], and Professor A.E. Verrill or Yale, are naturally much interested, and may be prevailed upon to visit."[12]

In fact, Webb had written to Dall - then curator of mollusks at the Smithsonian - urging him to "come down at once," but Dall's superiors denied him funds for travel. Neither would they pay to have the carcass shipped from Florida to Washington, D.C. Verrill likewise declined to make the journey from Connecticut. Taking rejection in stride, Webb cut several chunks from the globster for both malacologists.[13] His next letter to Dall, penned on 5 February 1897, read:

I made another excursion to the invertebrate and brought away specimens for you and for Dr. Verrill at Yale. I cut two pieces of the mantle and two pieces from the body and have put them in a solution of formalin for a few days before I send them to you. Although strange as it may seem to you, I could have packed them in salt and sent them to you at once although the creature has been lying on the beach for more than two months. And I think that both yourself and Dr. Verrill, while not doubting my measurements, have thought my account of the thickness of the muscular, or rather tendonous husk pretty large, so I am glad to send you the specimens and I will express them packed in salt in a day or two.[14]

Dr. Verrill could not wait. With the specimens in transit, he wrote two new articles on the huge octopus. One, published in February's *American Journal of Science,* formally named the new species *Octopus giganteus,* declaring: "It is possible that it may be related to *Cirroteuthis,* and in that case the two posterior stumps, looking like arms, may be the remains of the lateral fins, for they seem too far back for the arms, unless pulled out of position. On the other hand, they seem to be too far forward for fins. So that they are probably arms twisted out of their true position."[15]

At the same time, Verrill sent another piece to the *New York Herald,* appearing on 14 February 1897. It read, in part:

> The living weight of the creature was about eighteen or twenty tons. When living, it must have had enormous arms, each one a hundred feet of more in length, each as thick as the mast of a vessel, and armed with hundreds of saucer-shaped suckers, the largest of which would have been at least a foot in diameter....Its eyes would have been more than a foot in diameter. It would have carried ten or twelve gallons of ink in the ink bag. It could swim rapidly, without doubt, but its usual habit would be to crawl slowly over the bottom in deep water in pursuit of prey....We must reflect that wherever this creature had its home, there must be living hundreds or even thousands of others of its kind, probably of equal size, otherwise its race could not be kept up....[16]

With those opinions on record, Verrill prepared to examine the samples of tissue delivered by Webb - and they changed everything.

"Certainly Not a Cephalopod"

Verrill received the relics on 23 February 1897, and on 5 March wrote his third article on the St. Augustine globster, for the April issue of *American Naturalist.* "The supposition that it was an *Octopus*," he wrote, "was partly based upon its baglike form and partly upon the statements made to me that the stumps of large arms were attached to it at first. This last statement was certainly untrue." Without a shred of proof, Verrill essentially branded John Wilson a liar, calling his description of truncated arms "erroneous and entirely misleading." The bulk of the carcass, Verrill proposed, could be part of a whale, though "what part of any cetacean it might be is still an unsolved puzzle."[17]

In fact, a letter from Dr. Webb to the Smithsonian's Professor Dall, dated 17 March 1897, suggests that Dall influenced Verrill's change of heart. While no comments from Dall have been preserved, pro or con, Webb's letter read:

> As you already know, Prof. Verrill now says our strange creature cannot be a cephalopod and that he cannot say to what animal it belongs. I do not see how it can be any part of a cetacean as Prof. V. says you suggest. It is simply a great big bag and I do not see how it could be any part of a whale. Now that I have had it brought 6 miles up the beach it is out of the way of the tide and the drifting sand and will have a chance to cure or dry up somewhat. If it were not for the soft mass of the viscera which was so difficult to remove that we left it there would be but little odor. As it is there is no great amount.[18]

On 13 March, *The Tatler* weighed in to say: "Professor Verrill of Yale University, who recently decided that the curious something, supposed to be an octopus, was one, basing his decision on the descriptions sent, has now concluded, after examining a piece of it, that it could not possibly be an octopus, and he cannot decide what it is. One theory advanced is that it may be a portion of some inhabitant of the sea, long since extinct, that has been fast in an iceberg for centuries, and recently washed ashore here. Another theory is that it is a portion of a deep-sea monster that on coming too near the surface was attacked by a shark, who found it too tough for a breakfast. One thing is not determined, and that is, if we do not know what it is, we know what it is not.[19]

Presumably embarrassed, Dr. Verrill sat down on 19 March to write yet another article, this one for the April issue of the *American Journal of Science.* Its title - "The supposed great Octopus of Florida; certainly not a Cephalopod" - was emphatic, and Verrill sought to mask his former indecision by naming the Florida globster as part of a sperm whale. He wrote:

> The structure of the integument is more like that of the upper part of the head of a sperm whale than any other known to me, and as the obvious use is the same, it is most probable that the whole mass represents the upper

part of the head of such a whale, detached from the skull and jaw. It is evident, however, from the figures, that the shape is decidedly unlike the head of an ordinary sperm whale, for the latter is oblong, truncated and rather narrow in front, like the prow of a vessel with an angle at the upper front end, near which a single blow-hole is situated. No blow-hole has been discovered in the mass cast ashore.[20]

Having thus harpooned his own argument, Verrill urged his readers to "imagine a sperm whale with an abnormally large nose, due to disease or old age," admitting that "it seems hardly probable that another allied whale, with such a big nose, remains to be discovered. Notwithstanding these difficulties, my present opinion, that it came from the head of a creature like a sperm whale in structure, is the only one that seems plausible from the facts now ascertained."[21]

Waffling all the way - and stolidly ignoring Dr. Webb's reference to "viscera" inside the supposed giant whale's nose - Verrill risked becoming a laughingstock. Smithsonian curator Frederic Lucas did his best to help Verrill, perhaps referring to Dr. Dall's samples when he said, "The substance looks like blubber, and smells like blubber, and it is blubber, nothing more nor less." Still, the British journal *Natural Science* could not resist chiding Varrill: "The moral of this is that one should not attempt to describe specimens stranded on the coast of Florida, while sitting in one's study in Connecticut."[22]

Lost and Found

There matters rested for another sixty years. Today, no one recalls what became of the Florida globster, though it apparently served as a popular tourist attraction during early 1897. Sixty years later, Dr. Forrest Wood - senior scientist and consultant with the Ocean Sciences Department of the Naval Undersea Research and Development Laboratory in San Diego, California - was completing a study of octopus behavior at Florida's Marineland Research Laboratory (now the Whitney Laboratory for Marine Bioscience), when he discovered an old press clipping about the St. Augustine carcass. Intrigued, Wood launched a personal investigation and discovered, through a colleague at the University of Miami Marine Laboratory, that the Smithsonian Institution still possessed a large jar of tissue labeled *Octopus giganteus Verrill.*[23]

Wood next consulted fellow Marineland researcher and University of Florida professor

Joseph Gennaro Jr., who in turn contacted Harold Rehder, the Smithsonian's curator of mollusks. Invited to view the Smithsonian's samples, Gennaro flew to Washington and found "a glass container about the size of a milk can. Inside it was a murky mixture of cheesecloth, formalin (and I think some alcohol), and half a dozen large white masses of tough fibrous material, each about as large as a good-sized roast." Gennaro's efforts to excise a sample dulled the blades of four dissecting knives in turn, but he finally secured a portion of the globster for examination at his lab in Florida.[24]

Gennaro, a cell biologist, compared the St. Augustine tissue to known samples of whale,

ABOVE: DeWitt Webb with St. Augustine's globster, 1896.
BELOW: Another photo of *Octopus giganteus*, moved to higher ground

Professor Addison Verrill.

A sketch from 1896, depicting the Florida globster's arms.

squid, and octopus flesh. The results, as described in the March 1971 issue of *Natural History,* were astounding.

To my great dismay, no cellular material at all was discernible. Perhaps because the tissue mass had lain for so many days on the beach of St. Augustine, or perhaps because the formaldehyde or alcohol had had insufficient time to penetrate for adequate preservation, nothing of the original cellular architecture remained. I found, however, that my control samples, which had been properly prepared for histological analysis, also failed to show much cellular arrangement. But even more striking than the absence of cellular structure was the presence of distinctive patterns of connective tissue. Differences between contemporary octopus and squid tissue struck the eye immediately, and each was obviously different from the typical pattern of mammalian tissue.

It occurred to me that I might learn something by observing and comparing the connective tissue patterns of the specimens under polarized light. The highly ordered fiber protein molecules oriented in the plane of the section doubly refracted the light and showed up brightly, while those that were perpendicular appeared black.

Now differences between the contemporary squid and octopus samples became very clear. In the octopus, broad bands of fibers passed across the plane of the tissue and were separated by equally broad bands arranged in a perpendicular direction. In the squid there were narrower but also relatively broad bundles arranged in the plane of the section, separated by thin partitions of perpendicular fibers.

It seemed I had found a means to identify the mystery sample after all. I could distinguish between octopus and squid, and between them and mammals, which display a lacy network of connective tissue fibers.

After 75 years, the moment of truth was at hand. Viewing section after section of the St. Augustine samples, we decided at once, and beyond any doubt, that the sample was not whale blubber. Further, the connective tissue pattern was that of broad bands in the plane of the section with equally broad bands arranged perpendicularly, a structure similar to, if not identical with, that in my octopus sample.

The evidence appears unmistakable that the St. Augustine sea monster was in fact an octopus, but the implications are fantastic. Even though the sea presents us from time to time with strange and astonishing phenomena, the idea of a gigantic octopus, with arms 75 to 100 feet in length and about 18 inches in diameter at the base - a total spread of some 200 feet - is difficult to comprehend.[25]

Indeed. But Gennaro's verdict was not the last word on St. Augustine's globster.

Round Three

By 1985, Gennaro had transferred to New York University's Department of Biology, but he retained his interest in *Octopus giganteus.* That year, he subjected part of the remaining samples to comparative analyses of various amino acids, with control samples including mammalian collagen, tendon, and bone. Following completion of the tests, Gennaro submitted the results to Dr. Roy Mackal, a biochemist at the University of Chicago and a founding member - with Forrest Wood, Bernard Heuvelmans and others - of the International Society of Cryptozoology (ISC). Mackal, in turn, performed further analyses, employing control samples of tissue from *Architeuthis* and other squids, two species of octopus, a spotted dolphin, and a beluga whale. Mackal published their findings in the 1986 issue of the ISC's peer-reviewed journal *Cryptozoology.*[26]

Mackal wrote that "The *O. giganteus* tissue is almost pure collagen, which is precisely what one might expect for an aquatic invertebrate, such as a giant octopus, with a mass of 6,000 kg [13,200 lbs.] or more." Therefore, he concluded: "On the basis of Gennaro's histological studies and the present amino acid and Cu [copper] and Fe [iron] analyses, I conclude that, to the extent the preserved *O. giganteus* tissue is representative of the carcass washed ashore at St. Augustine, Florida, in November 1896, it was essentially a huge mass of collagenous protein. Certainly, the tissue was not blubber. I interpret these results as consistent with, and supportive of, Webb and Verrill's identification of the carcass as that of a gigantic cephalopod, probably an octopus, not referable to any known species."[27]

Another vote, then, for the giant octopus - but, once again, it would not be the final word.

Very Like a Whale

Most of DeWitt Webb's original globster photos were thought to be lost at the turn of the twentieth century, though copies of three from January 1897 survived at the Smithsonian Institution and were published for the first time by Roy Mackal, in 1980. The rest resurfaced in 1993, found by California resident Marjorie Blakoner in an album once owned by St. Augustine photographer Van Lockwood.[28] After studying the photos, author Richard Ellis wrote:

> Despite all the evidence, past and present, most investigators still insist that the St. Augustine monster was a whale. One has only to read the descriptive material and look at the pictures to realize that this is a misidentification. Indeed, if one examined the pictures with no information about the size of the object, there would be no question regarding the animal: It was an octopus.[29]

Photos meant nothing, however, to the next team of analysts, including Gerald Smith Jr. from Indiana University's School of Medicine and three members of the University of Maryland's Department of Zoology: Sidney Pierce, Timothy Maugel, and Eugenie Clark (another ISC board member). This time around, the researchers compared tissue from *O. giganteus* to tissue from a "blob" beached at Bermuda in 1988, flesh from an octopus mantle, tissue from a humpback whale, plus tendon and skin from a rat's tail. The samples were compared using both amino acid analysis and electron microscopy.[30]

The authors published their results in April 1995, and while they chose to lead with the 1897 "blubber" quote from Frederic Lucas, their findings were not so simple. First, they determined that both the Florida and Bermuda samples consisted of "almost pure collagen....Neither carcass is from a giant octopus nor any other invertebrate, but they are also not from the same species."

In fact, they declared, the Florida globster was a homeothermic (warm-blooded) creature, while Bermuda's globster was a poikilothermic (cold-blooded) vertebrate - i.e., some kind of fish. What "enormous warm-blooded vertebrate" washed ashore at St. Augustine? They conclude that it must have been "the remains of a whale, likely the entire skin." Rather smugly, authors close with an expression of "profound sadness at ruining a favorite legend."[31]

But had they done any such thing?

Leaving their treatment of the "Bermuda Blob" for dissection in Chapter 3, I join Richard Ellis in asking how a whale might lose its "entire skin" in one bag-like mass, washed ashore without a trace of flesh, muscle, or skeletal remains. (The latest study, once again, ignored DeWitt Webb's reference to viscera observed in 1897.) Whalers - then, as now - remove the flesh and blubber from their kills in strips. As with Professor Verrill's mythical big-nosed sperm whale, there is no case on record of any cetacean, dead or living, popping out of its skin like a molting insect. Ellis is clearly right in his assessment of the 1995 report: "we must conclude that the mysteries remain unsolved and the legend endures."[32]

Last Call - For Now

Authors Pierce, Smith and Maugel returned to the subject of globsters in June 2004, accompanied by colleagues Steven Massey and Nicholas Curtis from the University of South Florida, and Carlos Olavarría from the Centro de Estudios del Cuaternario Fuego-Patagonia y Antártica Punta Arenas, Chile. This time, they were concerned primarily with a globster stranded on the Chilean coast, in July 2003, but their remarks extend to the Florida globster, the aforementioned Bermuda Blob, an earlier Bermuda carcass, and a globster from Nantucket, Massachusetts.[33]

This time around, the team combined DNA analysis with electron microscopy, and despite repeated citations of their 1995 report - which deemed the 1988 Bermuda Blob the carcass of some unknown cold-blooded fish - they reached a radically different result. In fact, they declare: "It is clear now that all of these blobs of popular and cryptozoological interest are, in fact, the decomposed remains of large cetaceans.... The results, taken together, leave no doubt that all of the blobs examined here - St. Augustine, Bermuda 1, Bermuda 2, Tasmanian West Coast, Nantucket, and Chilean - represent the decomposed remains of great whales of varying species. Once again, to our disappointment, we have not found any evidence that any of the blobs are the remains of gigantic octopods, or sea monsters of unknown species."[34]

The authors offer no explanation for their apparent error in 1995, concerning the Bermuda remains - and, in fact, misrepresent those findings when they write: "Other relics such as the St. Augustine (Florida) Sea Monster and the Bermuda Blob are still described by some as the remains of a gigantic octopus (*Octopus giganteus*), even though A. E. Verrill - who named the St. Augustine specimen sight unseen - recanted his identification in favor of whale remains, and in spite of microscopic and biochemical analyses showing that they were nothing more than the collagenous matrix of whale blubber (Pierce *et al.,* 1995)."[35] As we have seen, their findings in the previous report said nothing of the kind concerning the Bermuda globster.

Still, the garbled findings satisfied the *New York Times,* which declared on 27 July 2004 that "[t]his does in fact appear to be the end of the great blob story, a tale that began in late 1896 near St. Augustine, Fla., when two boys found a gigantic lump of white, rubbery flesh, 21 feet long, 7 feet wide and weighing perhaps 7 tons."[36] But if the *Times,* was satisfied, ignoring contradictions in the two reports published ten years apart, skeptics acknowledge that unanswered questions still remain.

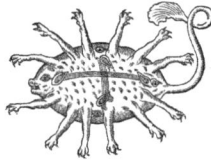

4.
A Globster Gallery

lobsters come in all shapes and sizes. The record of their strandings spans fifteen centuries, including every continent except Antarctica. While some are easily identified, others defy explanation - and, it may seem, common sense. Whether we take them seriously or dismiss them out of hand, the record stands. And it begins in the Far East.

Korea: 661/67 C.E.

Perhaps fittingly, out first story ranks among the most fantastic. It comes to us courtesy of linguist Marinus Van Der Sluus, from Korea's *Samguk Yusa* (*History of the Three Kingdoms*), written by the Buddhist monk Il-yeon (or Iryeon) in the late thirteenth century. While recording various Fortean events observed in Korea from its earliest times to the tenth century, Il-yeon reported one incident that fits - however roughly - into our catalog of beached wonders.

Specifically, he wrote (in Chapter 26, page 97) that "the body of a huge woman came floating on the sea south of Sabi-su" in either 661 or 667 C.E. According to ancient reports, "Her body was 73ft long, her feet 6ft long, and her mount of Venus 3ft long. Another story says her body was 18ft long." While contemporary witnesses regarded the floating corpse as "the goddess of the sea," Van Der Sluus suggests that it may have been a giant squid "or some other mollusc."[1] At this remove, speculation is fruitless.

Canton, China: 745

During this year, a giant "centipede" washed ashore at Canton (now Guangzhou), located on the Pearl River, some seventy-five miles northwest of Hong Kong. No further information is available, but if the creature was a marine denizen, it must have entered the Pearl from the South China Sea. Author Richard Muirhead notes an echo of the giant *Con Rit* (Vietnamese for "millipede") found at northern Vietnam's Along (or Halong) Bay in 1883.[2]

Scotland: 905

Koreans weren't alone in seeing giant humanoids at sea. According to the Irish *Annals of Innisfallen*, written in 905, "A woman was washed up on the shore of Alba in this year. Her length was two and 10 added to 20 times eight ft. [172 feet overall], her plaits were 16ft long, her fingers were 6ft, her nose 6ft and her body was as white as a swan or the foam of the wave." British historian Simon Young interprets "Alba" as a reference to Scotland, while noting that it might also refer to Britain as a whole. He rejects the giant squid solution offered for Korea's giantess by Marinus Van Der Sluus, proposing instead that Scotland's giant may have been "a very decomposed blue whale maybe, albino or otherwise."[3]

Lake Mjøsa, Norway: 1522

Our first report of a stranded lake monster was cited by Peter Costello in his 1974 book, *In Search of Lake Monsters*. Mjøsa ("the bright/shiny one") is Norway's largest lake, and boasts claims of "a serpent of incredible magnitude" dating from antiquity. Costello reports that an unknown cryptid was found at the lake and sold to unnamed German merchants. Late Norwegian researcher Erik Knatterud omits that detail, while claiming that locals killed the beast with "a volley of cross bow arrows."[4] It may thus belong more properly to Chapter 5, but in the absence of more detailed information, I present it here.

Baffin Island: 1577

In her *Atlas of the Mysterious in North America*, Rosemary Guiley reports that a dead "sea unicorn" was beached on Baffin Island, formerly part of Canada's Northwest Territories (now Nunavut), sometime in 1577. While offering no further details, she logically suggests that the creature may have been a narwhal (*Monodon monoceros*), a medium-sized cetacean distinguished by a long helical tusk protruding from its upper-left jaw.[5] (Some rare specimens boast double tusks.) There is no apparent reason to dispute Guiley's identification.

Wirral Peninsula: 1636

In February 1999, British correspondent Richard Holland alerted readers of *Fortean Times* to the cases of two forgotten globsters, subsequently expanded for the CFZ's online journal, *Still on the Track*. According to Holland's source - Christina Hole's *Traditions and Customs of Cheshire* (1970): "A very curious creature, called a Herring Hogg, was said to have been stranded on the shores of Wirral [North West England's Wirral Peninsula] in the spring of 1636. Sir John Bridgeman, Chief Justice of Chester, discovered it, when he was riding on the Lent Circuit. It was fifteen yards high, and twenty yards and one foot in length. Its voice was evidently powerful, for its cry could be heard six or seven miles away, and was 'so hideous that none dared come near it for some time.' We are not told what happened to it when the local people finally summoned up courage to approach." While surmising that "Herring Hogg is a generic name for any big sea mammal," Holland declined to offer a specific identification.[6]

Oaxaca, Mexico: 1648

British journalist and amateur historian Harold Wilkins described our next globster in his book *Secret Cities of Old South America* (1952), quoting a Spanish manuscript from 1650 that read:

In the year 1648, there appeared on the *playa* (beach) of Santa Maria del Mar, Oaxaca, a dreadful monster which, on the flood tide of the sea, was thrown up on the waves. Its bulk was great and appeared to the eye like a reef. The folk of the pueblo, 200 paces away from where it was, saw it at break of day, and were so terrified that they were on the point of quitting their houses. It moved and swayed slowly on the sands, and on the second day the motion was less. On the third day it was motionless. In eight days a bad smell arose form the huge carcase, and the folk saw birds swoop down from the sky and dogs began to eat the putrefying flesh. Convinced thereby that the monster was dead, the people plucked up courage to approach it. They found it to be 15 varas long (41.70 feet), and upon the sands, it exceeded two varas (5.56 feet) high. Its pelt was remarkable, or a red colour, like that of a cow. Its ears lacked folds (*cangilones*). It had two fore-feet, and its tail was like a pillar, being so oily and greasy, and stinking so much that not even the dogs could eat it.A shoulder-blade, shaped like a fan, was jointed, and a third of a vara in diameter (about eight and a half inches). Its rib was the width of an eighth (of a vara?) and two varas long (5.56 feet). The tail, or caudal extremity, reached to the shoulder blade and formed very singular buttocks.[7]

Wilkins further notes that a physician, one Don Juan Nepomuceno, confirmed the story, adding that a bone from the carcass once hung "facing east" in a window at "the library of the convent of Santo Domingo" - presumably Oaxaca's Church of Santo Domingo de Guzmán, which was once a monastery.[8]

Bernard Heuvelmans speculated that the creature may have been a stranded whale. Ben Roesch later agreed, in principle, while noting various inaccuracies found in Wilkins's text, together with his credulous acceptance of tales concerning Atlantis and the mythical lost city of El Dorado.[9]

Rogaland, Norway: 1720

Our next report is doubly strange, describing an unknown creature found at an untraceable location. We owe the account to Dutch biologist Antoon Oudemans, who quotes from the writings of Denmark's Bishop Pontoppidan.

Thorlack Thorlacksen has told me that in 1720 a Sea-Serpent had been shut up a whole week in a little inlet, in which it came with high tide through a narrow entrance of seven or eight fathoms deep, and that eight days afterwards, when it had left the inlet, a skin of a snake or serpent was found. One end of the skin had sunk into the water of the inlet, so that its length could not be made out, as the inlet was several fathoms deep, and the skin partly lay there. The other end of this skin was washed on the shore by the current, where everybody could see it; apparently, however, it could not be used, for it consisted of a soft slimy mass. Thorlacksen was a native of the harbour of Kobberveug.[10]

First, as to the location, Bernard Heuvelmans located "Kobberveug" in Norway's Rogaland county, but it appears that no such town exists, either in Rogaland or anywhere else in Norway. In all probability, Oudemans was referring to Kopervik, the largest city on Rogaland's island of Karmøy.[11]

In any case, Oudemans identified Thorlacksen's sea-serpent as a "gigantic calamary" - i.e., a giant squid, unknown to science in 1720. Heuvelmans and Ben Roesch concur in that judgment, dismissing the sea-serpent's "skin" as the decomposing corpse of *Architeuthis dux.*[12]

Nordfjord, Norway: Before 1753

Bishop Pontoppidan reports two more supposed sea-serpent strandings in Norway, the first occurring at Amunds Vaagen, Nordfjord, in Sogn og Fjordane county. Once again, the creature was beached at high tide - and once again, research reveals no present site called "Amunds Vaagen" anywhere in Nordfjord, a district spanning 1,658 square miles.[13] No further information is available today, but Roesch and Richard Ellis treat the report as a description of a stranded giant squid, while Heuvelmans leaves it with a cryptic question mark.[14]

Karmøy Island, Norway: Before 1753

Pontoppidan's last globster came ashore on "Karmen Island" - i.e., Karmøy - and was listed in the bishop's published work as yet another sea-serpent. As in the "Amunds Vaagen" case, authors Ellis and Roesch presume the creature was a giant squid, while Heuvelmans reserves judgment.[15]

Chester, England: 1782

Richard Holland included this case in his February 1999 letter to *Fortean Times,* quoting the *Chester Courant* to describe a carcass cast up by the River Dee near Chester, Cheshire, on 8 May 1782. As described in that article:

> The length of it is 25 feet; the girth proportionately large, though very unequal; it has two dorsal and six pectoral fins - two of the latter of a vast singular form, partaking of the nature of feet. The tail is perpendicular, of prodigious size and strength; there are five gills on each side. The mouth, when open to its extremity, is three feet wide; there are not any teeth, but a vast quantity of small, irregular sharp prominences, which are evidently given it for the purpose of comminuting its food, the orifice of the throat being astonishingly narrow for a creature of such magnitude.
>
> The upper and under jaws are each furnished with ten strong protuberant bones, horizontally placed, which meet when the mouth closed, in such a manner as to appear capable of breaking almost any substance. The eye is situated very near the mouth, and scarcely larger than that of an ox; the nose is hard and prominent; the whole body is covered with a very thin skin, and the weight of the fish is between four and five tons.

OPPOSITE: Bishop Pontoppidan, chronicler of sea-serpents.

ERICUS PONTOPPIDAN

S. Th. D. P. P. &
Episcopus Bergensis.

Huic Probitas, Huic Candor inest et Numinis Ardor
Ingeniumque Viro, teste vel invidia.

T. de Hofman
Assessor Summi Tribunalis sc.

We have been thus particular, as it is probable that some ingenious Naturalist may favour the public with the certain information of its real species.[16]

No answer was forthcoming at the time, but Holland notes that subsequent proposed identifications - including those of a basking sharp, a sperm whale, and a grampus (possibly referring either to Risso's dolphin [*Grampus griseus*] or the killer whale [*Orcinus orca*]) - have been rejected.[17] Clearly, no cetacean would have gills, and both proposed species have plainly visible teeth. Dismissal of the basking shark seems ill-advised, however, since the description of its mouth, nose and eye, plus the number of gill slits and dorsal fins, all conform to *Cetorhinus maximus*. The Cheshire carcass was, if anything, below average length for a basking shark, and while no shark has six pectoral fins, all possess paired pectoral and pelvic fins for stabilization while swimming. *C. maximus* also has a single anal fin, but the third pair of "pectoral fins" more likely represent the pelvic claspers of a male specimen - rolls of cartilage stiffened with calcium, using in mating.[18]

Stronsay, Orkney Islands: 1808

Thus far, we have dealt with little-known globsters, but our next qualifies as a superstar of the genre. The carcass was discovered on 26 September 1808 by farmer John Peace, while fishing near Rothiesholm Point on the southeastern coast of Stronsa (now Stronsay) Island. Peace first supposed it was a whale, stranded on offshore rocks and surrounded by flocks of seabirds, but closer examination proved it to be a strange creature beyond his knowledge. Another witness on shore, farmer George Sherar, watched Peace examine the carcass, then got a close look for himself when it beached on Stronsay proper ten days later.[19]

Peace was first to measure the carcass, declaring, declaring it to be "about fifty-four or fifty five feet in length." Sherar was next, using a foot-rule, and reported that it was "exactly fifty-five feet in length from the hole in the top of the skull...to the extremity of the tail." Carpenter Thomas Fotheringhame obtained the same result by measuring "from the junction of the head and neck, where there was the appearance of an ear, to the tail." The creature also had a mane, the witnesses agreed, with three pairs of fins or legs, each sporting five or six toes. Peace stated that one "fin or arm was edged all around from the body to the extremity of the toes, with a row of bristles about ten inches long."[20]

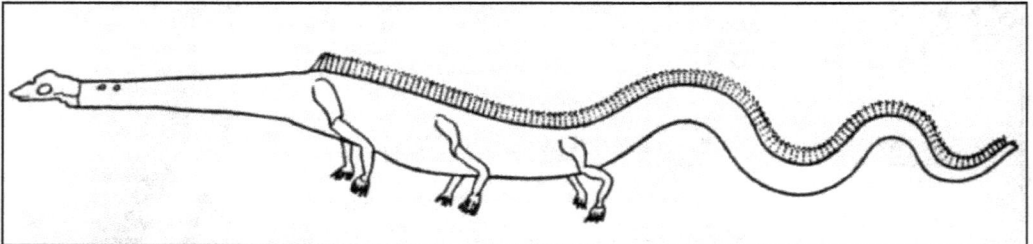

A sketch of the Stronsay carcass.

Sherar later provided a more detailed assessment of the carcass, in sworn testimony before Justice of the Peace Malcolm Laing, affirming -

> That the length of the neck was exactly fifteen feet, from the same hole to the beginning of the mane: That he measured also the circumference of the animal as accurately as he could, which was about ten feet, more or less; and the whole body, where the limbs were attached to it, was about the same circumference: That the lower jaw or mouth was awanting; but there was some substances or bones of the jaw remaining, when he first examined it, which are now away: That it had two holes on each side of the neck, beside the one on the back of the skull: That the mane or bristles were about fourteen inches in length each, of a silvery colour, and particularly luminous in the dark, before they were dried: That the upper part of the limbs, which answers to the shoulder-blade, was joined to the body like the shoulder-blade of a cow, forming part of the side: That a part of the tail was awanting, being incidentally broke off at the extremity; where the last joint of it was bare, was an inch and a half in breadth: That the bones were of a gristly nature, like those of a halibut, the back-bone excepted, which was the only solid one in the body: That the tail was quite flexible, turning in every direction, as he lifted it; and he supposes the neck to have been equally so from its appearance at the time: That there were either five or six toes on each paw, about nine inches long, and of a soft substance: that the toes were separate from each other, and not webbed, so far as he was able to observe; and that the paw was about half a foot each way, in length and breadth.[21]

Fotheringhame's sworn testimony in the same proceeding added that "the skin seemed to be elastic when compressed, and of a greyish colour, without any scales: it was rough to the feeling, on drawing the hand over it, towards the head; but was smooth as velvet when the hand was drawn towards the tail." He also noted that "a part of the bones of the lower jaw, resembling those of a dog, were remaining at that time, with some appearance of teeth, which were soft, and could be bent by the strength of the hand."[22]

The carcass had been disemboweled, with an apparent stomach dangling between the hindmost pair of limbs. Witness William Folsetter cut it open on a whim, later testifying "that it was about four feet long, and as thick as a firkin [a quarter of a standard full-size barrel], but flatter: That the membranes that formed the divisions, extended quite across the supposed stomach, and were about three sixteenths of an inch in thickness, and at the same distance from each other, and of the same substance, with the stomach itself: That the section of the stomach, after it was opened, had the appearance of the weaver's reed: That he opened about a fourth part of the supposed stomach which contained nothing but a reddish substance, like blood and water, and emitted a fetid smell."[23]

News of the "great Sea-Snake, lately cast in Orkney," reached Edinburgh's Wernerian Natural History Society on 19 November, with Malcolm Laing's promise to forward pieces of the carcass for examination. Bad weather delayed their delivery, but Dr. John Barclay had already examined the creature *in situ,* reporting back to the society on 14 January 1809, where he christened the animal *Halsydrus pontoppidani* ("Pontoppidan's water snake of the sea"). Affidavits from Stronsa were read before the society on 11 February, and April brought a claim from Rev. Donald Maclean that he had seen the carcass first, floating off the isle of Coll, in the Inner Hebrides, sometime in June 1808. Barclay's paper on the sea-serpent was published - with various drawings of the carcass - in 1811.[24]

Meanwhile, ichthyologist Sir Everard Home had examined a vertebra from the Stronsa globster, together with various sketches and eyewitness descriptions, declaring it to be a basking shark. Most modern authors agree, although the length as confidently described after multiple measurings exceeds the top acknowledged length for *C. maximus* by some thirty-eight percent.[25]

Murrumbidgee River, New South Wales: 1846

Our next case, from Down Under, involves the supposed skull of a *bunyip,* a freshwater cryptid described in various forms by Australian aborigines and Anglo-European settlers alike since 1789. Often compared to seals, swimming dogs, or larger mammals, *bunyips* are said to inhabit Outback lakes, rivers, and billabongs throughout eastern Australia and Tasmania.[26]

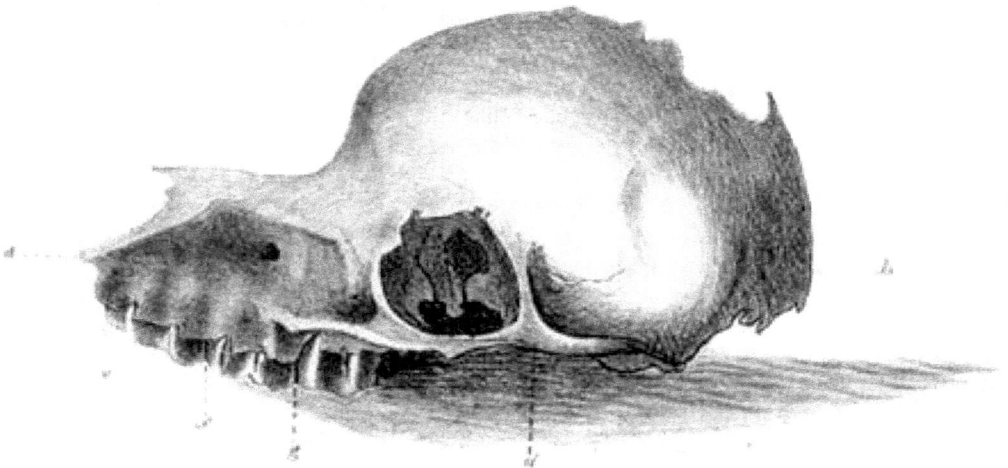

A drawing of the Murrumbidgee River skull.

In January 1846, settler Atholl Fletcher heard that native tribesmen had killed a *bunyip* along the Murrumbidgee River, and he visited the spot, retrieving a peculiar skull. As author Malcolm Smith describes it:

> The top of the cranium, the front of the snout, and the whole of the lower jaw was missing. According to [witness William] Hovell, the cranium measured 23 cm in length, from point A to point B [i.e., end-to-end]. To his untrained eye, the molars looked similar to those of an ox, and he had heard from the aborigines that the missing pieces of jaw would have supported enormous tusks.[27]

Fletcher subsequently took the skull to Melbourne, for examination by Dr. James Grant. Grant judged it to be the skull of a fetal or stillborn mammal, perhaps one of the camels that were introduced to Australia around 1840. Later still, the relic found its way to Sydney and the hands of British entomologist William Sharp Macleay. Macleay compared it to the skull of a deformed fetal colt, pulled from the Hawkesbury River in November 1841, and pronounced the *bunyip* a similar unfortunate creature. Sir Richard Owen, curator of London's Hunterian Museum, dissented after viewing sketches of the skull, which he declared to be a calf's. The skull was subsequently lost by staffers at the Colonial (now Australian) Museum, rendering further diagnosis impossible.[28]

Lagarfljót, Iceland: Before 1860[*]

In 1860, clergyman-novelist Sabine Baring-Gould introduced British readers to the *skrimsl*, a freshwater cryptid reported from various lakes in Iceland. After questioning eyewitnesses, Baring-Gould reported that the creature "measures 46 feet long, the head and neck are 6 feet, the body 22 feet and the tail 18 feet."[29] Aside from estimates and glimpses, Baring-Gould met a native scholar, one Dr. Hjaltalin, who had examined the supposed remains of a *skrimsl* at Lagarfljót, sometime earlier.

Lagarfljót - also called Lögurinn - is a lake located in eastern Iceland, near the town of Egilsstaðir. It is 15.5 miles long and 1.6 miles wide, with a maximum recorded depth of 364 feet. Dr. Hjaltalin described the *skrimsl* carcass found upon the lake's shore as a shapeless mass of flesh and bone which he could not identify, although the bones were "quite different from a whale's" or those of any other species known to Hjaltalin from local waters. No part of the carcass was preserved, and it appears that Dr. Hjaltalin published no summary of his findings.[30]

Fujian Province, China: 1863

On 5 January 1878 the British journal *Land and Water* published an article on sea-serpents, written by Dr. Andrew Wilson. The piece included an extract from the log of the schooner *Beaver*, penned by one Captain Boyle on 2 August 1863. At the time of the occurrence, Dr.

---

[*] In February 2012, the Icelandic national broadcaster, RÚV, published a video thought to show the Lagarfljót Worm swimming in snow-covered icy water, but it was later demonstrated to be making no progress and therefore probably an inanimate object moved by the rapid current.

Wilson described the ship's location as "the west coast of China, near Hamai, at Sungyce." Rightly noting that China has no west coast, Bernard Heuvelmans speculated that "Hamai is probably what is now Amoy [Xiamen to the Chinese, in Fujian Province]. Perhaps Sungyce was on a west-facing shore of the estuary there."[31] The exact location is untraceable today.

Wherever it occurred, in fact, Heuvelmans judged Captain Boyle's account "apparently authentic." The log's entry read:

> Well, here I am among the Sungyce Hakkas again. I came to anchor about twelve o'clock last night, about two miles out of this harbour. At half-past four o'clock this morning I went on shore with five young Chinese. The villages that are about three miles up the river were all in an uproar. I could not make out what was wrong with them - in fact I thought it was another fight. A little while longer, I saw them dragging at something, but what it was I could not tell...When I got a little closer...I saw that it was a great fish of some kind. He was not dead then...There were about 3,000 men and boys on the spot, every one with a lance, spear, knife, or chopper. More than half of these men were cutting and haggling at this monster. By the time I had been looking on, and walking around it, they managed to cut about forty feet off its tail or the small end of the monster, which is just the same as a snake's. I requested them to cut off its head, and said I would give them 500 cash to have a good look at the inside of its mouth. This was gladly accepted, while some were standing close to me as if they were out of wind at the hard work they had had with their choppers. I asked them how that fish came there. They told me that he came there at his own accord, and when on the sand made a fearful splashing and noise on the sand and water. At first every one of them were scared, until some of the fishermen ventured close to it, and called out that it was a large fish, and that it was theirs. This caused every one to run with whatever they could get to cut for himself. The fish ran on the bank at three o'clock....

> By this time the monster's head was cut off, but very much disfigured. I had them to draw it up the bank out of the water, and had the lower jaw cut off so as I could examine the inside of the mouth. I found the inside of the mouth to be just the same as a snake's, but it had three rows of soft teeth all as even as anything could be, and exactly the same size. They were movable, that is, I could move them towards the lip and back. At the entrance of the throat I found a strange sort of gridiron-shaped, tough substance, up and down. It was covered with a sort of reddish flesh, which causes me to think this monster of the deep lives on suckson [sic; suction?]. The snout was flat, the cheek or eye-brow stuck out about two and a-half feet - at least, two and a-half times the length of my boot. The skin was one and a-half inches thick only, but awful tough, of a dirty blue color. I should think there must have been many tons of barnacles on this monster. Where the barnacles were taken off there was a dirty white spot to be seen. As near as possible, it was twenty seven yards long. The head

is exactly like a snake's, but the eye was very like a hog's, till it was perfectly dead. My boots not being waterproof, and the sun being very hot, I was forced to leave, or I should have remained there until it was all cut up and weighed; but this I could not do. I have no idea of his weight. I left, and went ashore at a village close to the waterside, about two miles from the spot where the monster was being cut up. Here I found some tons of it upon the rocks being cut up into great junks [*sic*]. They were spoiling the bones by sawing them up as they were cutting the beef, as they call it.[31]

Heuvelmans speculated that the creature was "a big serpentine fish, a huge unknown selachian" - shark - while noting that "a shark's tail is not single like a snake's, but with two fins." Even so, the upper caudal fin in some species, such as the thresher and tiger, is so much longer than the lower that it might, in decomposed or mutilated specimens, be taken for a pointed tail.[32]

That said, and despite the triple row of teeth common to sharks, Heuvelmans offered an unknown giant eel as an alternative suspect. Ben Roesch rejects that suggestion, while conceding that it is "conceivably within the realm of possibility." From Captain Boyle's description of the monster's bones, tough blue skin and heavy coat of barnacles, Roesch concludes that the beast was most likely some species of baleen whale (suborder *Mysticeti*). The final truth, as Roesch concedes, will never be known.[33]

St. Margaret's Bay, Nova Scotia: Before 1864

In his 1965 tally of sea-serpent strandings and captures, Bernard Heuvelmans notes that "a lady" found something or other, somewhere along the forty-odd miles of coastline surrounding St. Margaret's Bay, at some unknown time prior to 1864. He offered no further data and declined to speculate on the creature's identity. Three decades later, Ben Roesch drew a blank on the case, while opining that the globster "was probably the carcass of a basking shark or whale."[34]

Loch Ness, Scotland: 1868

On 8 October 1868, the *Inverness Courier* reported that a carcass had been found near the village of Abriachan, where the Kilianan stream feeds Loch Ness. It proved to be a northern bottlenose whale (*Hyperoodon ampullatus*), left on the shoreline to dazzle "primitive" local inhabitants.[35]

Monhegan Island, Maine: 1880

On 5 June 1880, while sailing past Monhegan Island on the coast of Lincoln County, Maine, Captain M.D. Ingalls spied something floating on the surface near his ship, the *Chalcedony*. As he described the object:

> It was dead and floated on the water, with its belly, of a dirty brown color, up. Its head was at least 20 feet long, and about 10 feet through at the thickest point. About midway of the body, which was, I should guess, about 40 feet long, were two fins, of a

clear white, each about 12 feet in length. The body seemed to taper from the back of the head down to the size of a small log, distinct from the whale tribe, as the end had nothing that looked like a fluke. The shape of the creature's head was more like a tierce [a 42-gallon keg] than anything I can liken it to. I have seen almost all kinds of shapes that can be found in these waters, but never saw the like of this before.[36]

Despite the captain's confidence that it was not a cetacean, both Heuvelmans and Roesch disagree, casting their votes for a dead humpback whale (*Megaptera novaeangliae*). The dimensions cited would make it a relatively small adult, and the missing flukes may be explained by predation or decomposition.[37]

Marlboro, New Jersey: 1881

On 14 December 1881 the *New York Times* published a remarkable story from Marlboro, New Jersey. It read:

> The finding of the remains of the large sea serpent in the marl pit of O.C. Herbert at this place last week was supplemented yesterday by the discovery of the remains of another one. The find of yesterday was somewhat decomposed, and only two large tusks and portions of the jawbone of the reptile were found preserved, the other bones crumbling to pieces when exposed to the air. The bones found last week are all well preserved, and the tusks are remarkable for their size and their fine natural polish. Prof. Samuel Lockwood gives the following description of the reptile: "It was a monster of remarkable bulk. It had two paddles well forward, the body being short and stout. The bones of the paddles, from their size and solidity, indicate extraordinary propelling power. The tail was stout, long, and serpentine, but a little flattish, thus affording great aid in propulsion by a sculling movement. The neck was long, and yet thick enough to support the head high out of the water while the monster was engaged in devouring its prey. The huge jaws were armed with tusks which were more formidable than those of the crocodile. The lower jaw was very singular in structure, and had a joint like an elbow. In the act of swallowing the reptile could enlarge its gullet by means of this elbow joint. The act of swallowing was necessarily slow, and the reptile no doubt would have had great trouble in retaining in its mouth its struggling prey if it had not been for a supplementary jaw which was used as a grapnel. This was armed with small teeth, which were curved in shape and very sharp. As the large jaws, with their great tusks, were being opened so as to obtain a new hold the little grapnel jaw held the struggling prey fast, and the movements alternated until the fish or other prey was forced down the great throat." The only bones of this monster reptile of the antediluvian age known are those found in the marl pits of Mr. Herbert, and no name has yet been found for it by the scientists.[38]

A follow-up article, dated 21 December, noted that O.C. Herbert had refused all offers for the sea-serpent remains, while indicating that he might sell them to New Jersey's state geologist, Professor George Cook, for display at the capital museum in Trenton.[39]

While a casual read might suggest the discovery of two - count 'em, two! - epic globsters, common sense deflates the initial rush of excitement. First, Marlboro lies nearly ten miles inland from the New Jersey's coastline, too far for even the largest sea monster to stretch. Furthermore, New Jersey's marl pits were laid down during the upper Cretaceous period, 100 to 140 million years ago.[40] Clearly, whatever the nameless Marlboro beasts may have been, they were fossil remains.

Moeraki, New Zealand: 1883
In March 1883 a supposed sea-serpent washed ashore at Moeraki, a small fishing village on the east coast of New Zealand's South Island. Heuvelmans places it "near Otago," which is in fact New Zealand's second-largest region, spanning twelve thousand square miles. A brief description of the carcass made it 12.5 feet long, 15.25 inches deep, and 3.5 inches thick. While no conclusive evidence remains, Heuvelmans was probably correct in calling it an oarfish (*Regalecus glesne*). Described in many accounts as Earth's longest fish, *R. glesne* - alias "King of the Herrings" - is known to exceed thirty-five feet, and some published reports claim a confirmed record length of fifty-six feet.[41]

Queensland, Australia: 1883
On 19 March 1883 the *New Zealand Times* reported the discovery of a forty-foot globster beached on the east coast of Queensland, Australia. The carcass, including an "enormous" hip bone, were transported to Rockhampton. According to the *Times,* "There are the remains of what must have been an enormous snout, 8 feet long, in which the respiratory passage was yet traceable." Charles Fort, collecting the item for inclusion in his fourth book of oddities, *Lo!* (1931), opined that: "These could not have been the remains of a beaked whale. Whatever hip bones a cetacean has are only vestigial structures. In a sperm whale, 55 feet long, the hip bones are detached and atrophied relics of former uses, each about one foot long."[42]

Fort was correct, but Bernard Heuvelmans, nonetheless, offers a tentative I.D. of the carcass as a whale's. Ben Roesch agrees, on rather curious grounds, writing: "As for the sea serpent's enormous 'hip bone,' it is hard to say what part of a whale might be responsible. It is possible that the observers were mistaking other bones (such as the ribs or vertebrae for a hip bone. It should also be noted that very few aquatic vertebrates have very large hip bones (the pelvic girdle)."[43]

Of course, mistakes are always "possible," but none is demonstrated in this case. And if "very few" aquatic vertebrates sport large pelvic bones, it might be instructive to note which ones *do.* Alas, Roesch offers no clue - and his analysis fails to allow for any possibility that the Queensland carcass represented a new species. Finally, intent on proving his case, Roesch asks:

What are we to make of the trunked sea serpent? ... Certainly decomposed whales can take on very strange forms, and [Michael] Bright notes that "humpback whales, which migrate along the outer edge of the Great Barrier Reef, sometimes strand in this area." Possibly then, the Queensland trunked monster was really a dead humpback, which had been decomposing at sea for some time before being tossed up on shore. The "trunk" may not have been a long nose (as in elephants), but instead a result of the whale's skin and blubber decomposing and peeling off. This could form rolls of rotting flesh that might be interpreted as a "trunk."[44]

Perhaps, indeed - except for the fact that no one mentioned a "trunk" until Roesch penned his analysis of the Queensland globster in 1998.

Brungle Creek, New South Wales: 1883
Australia yielded another mystery carcass six months after the Queensland globster, chronicled once again by Charles Fort. This time, his source was the *Adelaide Observer* of 15 September 1883, reporting that a local resident named Hoad had found a most peculiar corpse on the bank of Brungle Creek, in the Riverina region of southwestern New South Wales. A veritable *bunyip,* it was headless, but the body that remained was "pig-like...with an appendage that curved inward, like the tail of a lobster."[45] It was not preserved for scientific study, and remains unidentified today.

Hongay, Vietnam: 1883
Our next case, one of the most enigmatic on record, was initially reported by Dr. A. Krempf, director of the Oceanographic and Fisheries Service of Indo-China [now Vietnam], in a letter to Professor Abel Gruvel, then director of the marine laboratory at a museum in the Brittany region of northwestern France. Krempf wrote:

Here is some information which although it smacks of the marvellous, cannot fail to interest you. I received it at sea from the coxswain of a Customs launch, a 56-year-old native called Tran Van Con.

38 years ago [i.e., in 1883], this Annamite saw and touched the so-called sea-serpent. Here is his account, faithfully translated: the animal was washed up and dead: it was a carcase in a very advanced state of putrefaction. The head was gone. The body alone was 60 feet long by 3 feet wide.

The animal was formed of successive segments almost all alike one another. Each segment was 2 feet long and 3 feet wide and had a pair of appendages 2 feet 4 inches long.

The teguments were of a remarkable consistency and rang like sheet-metal when hit with a stick. The colour of this tegumentary envelope was dark brown on the dorsal surface and light yellow on the ventral surface.

The stench that arose from this prodigious animal was such that even the Annamites would not go near it, and it was decided to tow the remains out to sea and sink them.

The name given to this animal by my informant is *con rit,* or "millipede." It is thus, according

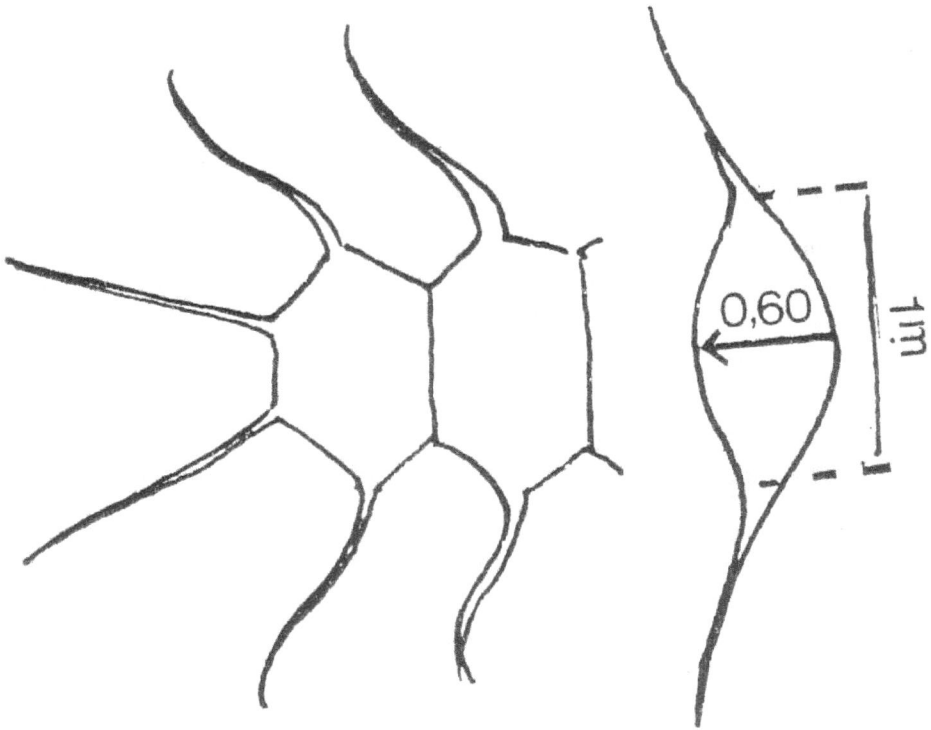

Contemporary sketches of the Vietnamese *con rit.*

to its name and from all the description that I have given you, an Arthropod...unless it is all a dream, and certainly it is very detailed, and as another theory about the sea-serpent can do no harm. I have thought fit to send you this information, but ask you to await further details before doing anything about it.

The event occurred at Hongay in Along Bay 38 years ago, and I had confirmation of it from a 30-year-old Chinese who had heard the tale from his father.[46]

While questioning Tran Van Con, Dr. Krempf saw a horseshoe crab lying on the beach nearby and struck it with a stick, asking the coxswain if it sounded like the noise produced by hitting the *con rit.* Tran Van Con replied, "Exactly."[47]

Bernard Heuvelmans, discussing the case at length, ultimately failed to propose a viable *con rit* candidate, despite suggesting that it might represent some unknown armored species of prehistoric cetacean (suborder *Archaeoceti*). Karl Shuker cast his cautious vote for a giant, unidentified crustacean. Ben Roesch was more prosaic, suggesting that the *con rit* may have been a dead whale's fleshless spinal column - or perhaps that of an oarfish.[48] All we may be certain of, today, is that nobody knows.

New River Inlet, Florida: 1885

Dr. J.B. Holder, a chronicler of Florida history, introduced readers of *Century Magazine* to our next globster in June 1892. According to his article -

> In the spring of 1885 the Rev. Mr. Gordon of Milwaukee, President of the United States Humane Society, chanced to visit, in the course of his duties, a remote and obscure portion of the Atlantic shores of Florida.
>
> While lying at anchor in the New River Inlet the flukes of the anchor became foul with what proved to be a carcass of considerable length. Mr. Gordon quickly observed that it was a vertebrate, and at first thought it was probably a cetacean. But, on examination, it was seen to have features more suggestive of saurians. Its total length was forty-two feet. Its girth was six. The head was absent; two flippers, or fore-limbs, were noticed, and a somewhat slender neck, which measured six feet in length. The carcass was in a state of decomposition; the abdomen was open, and the intestines protruded.
>
> The striking slenderness of the thorax as compared to the great length of the body and tail very naturally suggested to Mr. Gordon, whose reading served him well, the form of some of the great saurians whose bones have so frequently been found in several locations along the Atlantic coast. No cetacean known to science has such a slender body and such a well-marked and slender neck....Appreciating the great importance of securing the entire carcase, Mr. Gordon had it hauled above the high-water mark, and took all possible precautions to preserve the bones until they could be removed....He counted without the possible treacherous hurricane; the

The New River Inlet carcass.

waters of the "still-vexed Bermoothes," envious of their own, recalled the strange waif.[49]

Before proceeding, we must note one error in Holder's account - specifically, the reference to events occurring in spring 1885. That year's hurricane season featured eight storms between 7 August and 13 October, with the first to pass Florida's Atlantic coast occurring on 21 September. Another struck Florida's eastern coast on 10 October. Clearly, neither date qualifies as "spring."[50]

The location of the globster's discovery is also problematic. Modern America's only New River Inlet is located on North Carolina's Onslow Bay, but Florida claims two New Rivers: one in Broward County, the other serving as a border between Bradford and Union Counties. The first spills into the Atlantic at Port Everglades, while the latter is a tributary of the Santa Fe and does not reach the sea. (Ben Roesch, meanwhile, locates New River Inlet near Fort Pierce, which has no location of that name.) We must assume, therefore, that Rev. Gordon found his beastie somewhere in the neighborhood of modern-day Fort Lauderdale, known in the nineteenth century as the New River Settlement.[51]

As to *what* the creature was, Heuvelmans dismissed basking sharks on the theory that Florida's waters are too warm, but noted historical strandings of whale sharks along the state's coast, finally casting his vote for an unidentified "large selachian." Roesch agrees in principle, while noting that basking sharks have been found off the Florida coast - a fact confirmed by the Florida Museum of Natural History.[52]

Cape May, New Jersey: 1887

Two years later, and some 950 miles to the north, another globster surfaced somewhere along southern New Jersey's Cape May Peninsula. Heuvelmans recorded its discovery in an appendix to his epic work, but made no mention of it in his text. A list of sources for strandings omitted from further discussion cites a November 1887 issue of the long-defunct *Boston Courier,* otherwise undated. Ben Roesch, reviewing the Father of Cryptozoology's work in 1998, could not locate the article in question. "Until I am able to do so," he wrote, "this carcass remains a literal unknown."[53]

But no longer.

While researching *Globsters*, I obtained a copy of the long-missing report, with generous assistance from Henry Scannell, curator of Reference and Information Services at the Boston Public Library. The story, headlined "The Stranded Sea Serpent," appeared on page one of the *Boston Courier*, on 6 November 1887. It reads:

> The sea serpent business is practically ruined. A genuine specimen of the tribe has been washed ashore at Cape May, and the mystery is forever dispelled. Who will be greatly excited by a glimpse of a creature which will now be set down in the tables of the naturalists, and ticked with a long and learned name? When there were grave

doubts whether the sea serpent really existed, then indeed there was a certain amount of pleasure and excitement in running across a monster declared by half the world to be fabulous; but now the whole business is reduced to the level of the commonplaces.

Now that the serpent is in the hands of scientists, moreover, they have measured him, and it is painful to see how far short he falls of his supposed length. What are a paltry dozen feet beside the forty or fifty with which tradition had endowed him? Who can ever again really respect a beast that has fallen so absurdly short of its reputation? Even its big mouth and thick hair can do very little toward making up for this fatal deficiency of extent, and unless a bigger specimen can be forwarded from beneath the seawater immediately the fame of the race is irretrievably injured. Proprietors of seaside resorts which depend for success upon the reported presence of the sea serpent off their shores should send a communication at once to the king of the clan to lay the case before him, and to see if a bigger specimen cannot be induced to sacrifice himself for the good of his kind by coming to be measured. With a little diplomacy much may yet be done to save the fading fame of the sea serpent, but it must be done quickly, before the specimen ashore at Cape May shall have been generally received as typical.

No "long and learned name" was forthcoming, but the facetious article provides certain basic information about the Cape May globster - specifically, its twelve-foot length, plus an allusion to its "big mouth and thick hair." It is somewhat surprising that Dr. Heuvelmans, with said article in hand, did not suggest a possible identify for the stranded creature. He made do with less in other cases, including five between 1808 and 1896 wherein he proposed the existence of large unknown sharks.[54]

The Cape May specimen's "big mouth" instantly suggests that common globster candidate, the basking shark, well known to ichthyologists since 1765. A possible alternate suspect might be the megamouth shark, unrecognized by science until 1976. Both exceed the relatively modest twelve-foot length of Cape May's "serpent," with record lengths of forty feet and eighteen feet, respectively.

But what of the monster's "thick hair"? Fish have none, but as a shark's skin decomposes, the underlying muscle fibers may protrude, resembling hair that varies in color from white to red. Such was the case, we're told, with Scotland's fifty-five-foot Stronsa beast of 1808, which Heuvelmans confidently labeled a "giant basking shark or unknown shark."[55] Again, his hesitation in identifying Cape May's specimen may be the greatest mystery involved in the affair.

And yet ....

Who were the unnamed scientists described as measuring the carcass and removing it for further study? Where did they take the carcass? Did they ever reach a final verdict on the stranded beast's identity? If so, that finding never rated mention in the *Boston Courier* or any other East Coast newspaper so far discovered, much less in the august *New York Times.* Perhaps, therefore, some vestige of the mystery endures.

Connemara, Ireland: ca. 1888

Details are sparse concerning the accidental death of a freshwater cryptid - or "horse-eel" - in the Connemara district of western Ireland. Author F.W. "Ted" Holliday reported the event in his book *The Dragon and the Disc* (1973), relating the case of a large eel-like creature that became trapped in a culvert linking Crolan Lough to Lough Derrylea, west of Letterfrack (nine miles northeast of Clifden, in County Galway). While first seen alive, the creature subsequently died and smelled so bad that locals "didn't bother going near it and it stayed and it just melted away."[56]

Coffin Bay, South Australia: 1891

Our next case begins in journalistic confusion and ends, after a fashion, with latter-day claims of a hoax. The initial flap began with a telegram from Australia, received in London by Dalziel's Press Agency. It read, sans anything resembling punctuation: INFLUENZA EXTENSIVELY PREVALENT WALES VICTORIA NUMEROUS DEATHS BISHOP ADELAIDE FOUND DEAD SEA SERPENT SIXTY FEET COFFIN BAY. The agency, as it later explained, judged the last six words to be a separate sentence, deleting them as "unsuitable" before submitting the rest to *The Times* - which published the remainder on 6 November 1891.[57]

The competing *Saturday Review* gleefully lampooned *The Times,* noting that "Bishops are not generally 'found dead,' but die - when they cannot help it - in a decorous manner, and in the presence of witnesses." As for the deleted cryptid, "Did they suppose that a sea-serpent had come within sixty feet of Coffin Bay, or had devastated sixty feet of the shore, or that a sea-serpent with sixty feet had invaded that cheerfully-named locality?"[58]

In fact, *The Times* acknowledged on 9 November that Adelaide's bishop - the Very Reverend G.W. Kennion - was very much alive. A subsequent edition, published on 16 December 1891, clarified the embarrassing matter.

Yesterday's Australian mail brought news of the finding by the Bishop of Adelaide of the carcase of a sea serpent at Avoid Point, near Coffin Bay, South Australia. The Bishop, in writing to an Adelaide friend, states that while riding along the sea beach he came across a dead sea serpent about 60 ft. in length. It had a head 5 ft. long, like that of an immense snake, with two blow holes on the top. There were no teeth in the jaws. The body was round, and the tail resembled that of a whale. The Bishop described his "find" as the most peculiar animal he has ever seen.[59]

Thus was the discovery authenticated - at least, until 1892, when Antoon Oudemans refuted it in his book *The Great Sea Serpent.* According to Oudemans, one Gilbert Bogle of Newcastle-upon-Tyne, England, wrote to Bishop Kennion following the *Times* report of 16 December, and received an answer stating that the tale was false. No further explanation was forthcoming, though Ben Roesch surmises that persons unknown fabricated the later story as a joke, inspired by the original November telegram. That said, baleen whales *do* have twin blowholes - the only cetaceans so endowed - so there remains an outside possibility that something was indeed beached at Coffin Bay, with or without Bishop Kennion in attendance.[60]

Kirkwall, Orkney Islands: 1894

Bernard Heuvelmans reports that a globster was stranded near Orkney's capital and largest town, sometime in 1894, then dismisses it as a basking shark without offering any details. His short list of sources for strandings not mentioned in the text, likewise, offers nothing concerning this case. My search for further information on the incident proved fruitless.[61]

Tuamotu Archipelago: 1890s

Another case receiving short shrift from Heuvelmans occurred sometime during the nineteenth century's last decade, somewhere amid the Tuamotus - a myriad of islands and atolls in French Polynesia. After relating his encounter with a live sea-serpent in the area, date unknown, British traveler Ernest Davies added that: "A short time afterwards I heard from a reliable source that a sea-serpent had been cast up by a tidal wave on a reef some distance off. It measured fifty-three feet in length, and had a girth of twelve feet."[62]

While omitting the case from his list of strandings, Heuvelmans wrote, "I have not been able to find out anything about this washed-up sea-serpent. But, considering how many carcases have proved to be known whales or sharks, little has probably been lost."[63] And there the matter rests, presumably forever.

Suwarrow, Cook Islands: 1899

In February 1899, while en route to Australia, the British merchant steamer *Emu* stopped at Suwarrow, midway between Samoa and Tahiti in the South Pacific. On arrival, the sailors learned from natives that a large "devil-fish" had been stranded nearby. A team of *Emu* crewmen found the decomposing carcass, measuring sixty feet long, brownish colored, and covered with hair. The creature's skull was three feet long and resembled a horse's, except for two "tusks" on the lower jaw. Its largest ribs were thirty inches long, and its vertebrae measured four inches in diameter. Native tribesmen described two pectoral "flappers," present when the carcass came ashore but subsequently washed away. Observers pegged its weight at sixty tons. Oddly, a later story in the *Sydney Morning Herald* proclaimed the beast to have two heads.[64]

Despite the creature's rancid smell, sailors retrieved its skull for examination at Sydney's Australian Museum, where Dr. Edgar Waite identified its "tusks" as the characteristic teeth of a male beaked whale (family *Ziphiidae*). Waite did not venture to suggest a species, and with good reason: the largest known ziphiid - Baird's beaked whale - claims a record length of thirty-nine feet for a male specimen, forty-two feet for a female. Stranger still, its range is confined to the North Pacific.[65]

A relative of Baird's beaked whale - Arnoux's beaked whale - does occupy the Southern

Hemisphere, but it is smaller yet, with published estimates of its maximum length ranging from thirty-two to thirty-nine feet.[66] Whatever came ashore on Suwarrow, the carcass was clearly no common beaked whale.

Caledonian Canal, Scotland: ca. 1899-1900

The Caledonian Canal divides the Scotland's Great Glen, connecting Corpach (near Fort William) on the south to Inverness and Moray Firth in the north. Roughly one-third of its sixty-two-mile course in manmade, the remainder consisting of four natural lakes. From south to north they are Loch Dochfour, Loch Lochy, Loch Oich and Loch Ness. The canal also features twenty-nine locks, four aqueducts, and ten bridges.[66]

Mike Dash offers a third-hand account of a carcass found at Corpach roughly 111 years ago, relayed to him by "Irish monster-hunter Captain Lionel Leslie, from a Mrs. Cameron." According to the lady, writing "probably early in the 1960s," an unknown creature was found dead "in the Corpach canal-locks when these were drained at the end of the last century." In one article, Dash dates that event from 1899; in another, from 1900. His second take on the case places Corpach vaguely "south of Loch Ness," while claiming that "[i]t was assumed to have come from the loch." That assumption is odd, to say the least, since Corpach stands on the north shore of Loch Linnhe, thirty-four miles south of Loch Ness, and is thereby linked to the Firth of Lorne - which in turn connects to the Gulf of Corryvreckan and the North Atlantic beyond.[67]

As to what the creature may have been, Mrs. Campbell wrote: "In appearance it resembled an eel but was much larger than any eel ever seen and it had a long mane."[68] We know nothing beyond that, today, but large eels - with or without "manes" - are more likely to reach Corpach's sea-locks from the ocean, than by traveling southward along the canal.

Newport Beach, California: 1901

The present-day city of Newport Beach, in California's Orange County, did not exist in 1901, although the captain of the steamer *El Vaquero* had named a local lagoon "Newport" in 1871, and surrounding Orange County was created eighteen years later. Local resident Horatio Forgy - an Ohio Civil War veteran, later a director of Orange County's Santa Ana Hospital - may have had politics in mind on 22 February 1901, as he strolled along Newport Beach, but if so, they were soon forgotten at sight of a carcass stretched out on the sand. His description was sketchy at best, stating that the beast measured "twenty-one feet and some inches long," with its weight estimated at "about 500 or 600 pounds."[69]

Thankfully, at least one photograph was snapped of the carcass, later reproduced by Dr. Heuvelmans in his 1965 survey of sea-serpent sightings. Two days after Forgy discovered the carcass, the *Los Angeles Times* reported that "[h]undreds of people from this city wended their way to Newport Beach today to get a glimpse of the deep-sea monster which has washed ashore there." By that time, the beast had been identified as an oarfish, thus allaying any fears of a sea-monster invasion.[70]

Iceland: 1903

Our next case is another of those passed over by Heuvelmans with the briefest mention and lacking a source. His entry reads, in its entirety: "1903 (?Aug) - Iceland - *Indian* = Oarfish."[71] Both the incident and ship remained untraceable during my research for *Globsters,* leaving us with Heuvelmans's confident verdict, presumably based on something.

Strait of Dover, England: 1906

This incident occurred off Dungeness, a headland on the coast of Kent. A merchant ship, the *Tropper,* was passing on 20 March 1906, when its captain - Rathbone by name - saw "a carcase 50 feet long with small ears and white stripes which he took to be a sea-serpent." Based on that brief description, Dr. Heuvelmans could offer no suggestion of identity, leaving the case tagged with another nagging question mark.[72]

Ben Roesch felt more confident, thirty years later. While leaving the creature's "small ears" unexplained, he suggested that its visible stripes might represent the ventral grooves found on rorquals, the largest baleen whales (family *Balaenopteridae*). The largest member of the family - and Earth's largest known animal - is the blue whale (*Balaenoptera musculus*), which approach 100 feet in length. "Then again," Roesch admits, "maybe the *Tropper's* carcass represents a bona fide sea serpent. In the end, though, the details are simply too scarce to make any solid conclusions."[73]

And there the matter must remain, another mystery.

Sumbawa, Indonesia: 1906

Sumbawa is an island in the Lesser Sunda chain, flanked by Lombok to the west and Flores to the east. Sometime in 1906 it was the scene of yet another carcass stranding listed by Dr. Heuvelmans without elaboration or sources, stating simply that the creature was an oarfish.[74] While that judgment was correct, my research for *Globsters* revealed that the fish was not stranded or captured at all.

British naturalist Frederic Wood Jones (1879-1954), writing in *The Fishes of the Indo-Australian Archipelago*, reported that the incident occurred on 28 October 1906, thirty miles south of Sumbawa in the Indian Ocean. Members of a ship's crew saw a "long and very beautiful fish came to the surface at the ship's bow. Baited rigs were thrown to it, but it took no notice of them." One witness remarked that with "its vivid red crest and dorsal fin, scarlet streamers on its sides, and blue of its head and intense shine of silver on its body, it was probably the most beautiful creature I've ever seen."[75] Thus, clearly, Heuvelmans was right - and wrong, at the same time.

North Atlantic Ocean: 1908

Charles Fort reported our next case, in the aforementioned *Lo!* That entry read:

> In looking over the London *Daily News,* I came upon an item. Trawlers of
> the steamship *Balmedic* had brought to Grimsby the skull of an unknown

monster, dredged up in the Atlantic, north of Scotland (*Daily News,* June 26, 1908). The size of the skull indicated an animal the size of an elephant, and it was in "a wonderful state of preservation." It was unlike the skull of any cetacean, having eye sockets a foot across. From the jaws hung a leathery tongue, three feet long. I found, in the *Grimsby Telegraph,* June 29th, a reproduction of a photograph of this skull, with the long tongue hanging from the beak-like jaws. I made a sketch of the skull, as pictured, and sent it with a description to the British Museum (Natural History). I received an answer from Mr. W.P. Pycraft, who wrote that he had never seen any animal with such a skull - "and I have seen a good many!" It is just possible that nobody else has ever seen anything much resembling a sketch that I'd make of anything, but that has nothing to do with descriptions of the tongue. According to Mr. Pycraft no known cetacean has such a tongue.[76]

In fact, the ship's name given in the *Daily News* of 26 June 1908 was *Balmedie,* not *Balmedic.* Fort's correspondent was William Plane Pycraft (1868-1942), renowned British osteologist, a curator at London's Natural History Museum from 1907, and the author of ten books on various animal species published between 1901 and 1934.[77] As an expert in animal bones, he would seem qualified to judge the *Balmedie* skull.

Nonetheless, while citing *Lo!* and both newspaper articles (and repeating Fort's error in naming the *Balmedie*), Bernard Heuvelmans dismissed the find as a whale's skull, declining to mention the case in his text. Ben Roesch concurred, thirty years later, writing: "I was unable to obtain the original reports regarding the *Balmedic*'s skull [*sic*], and thus could not examine the photographs. However, despite the possible doubt regarding a whale identity brought up by Fort, I remain convinced that the *Balmedic*'s skull was indeed that of a whale."[78]

Roesch also notes, correctly, that some whales possess very prominent tongues. A humpback whale stranded on Alaska's Admiralty Island in July 2007 had a tongue swollen to the size of a small compact car, while a living blue whale's tongue may weigh three tons.[79]

Okanagan Lake, British Columbia: 1914

Lake monsters strand themselves less frequently than creatures from the sea, but it happens. One such case, perhaps, occurred in 1914 at Okanagan Lake, rumored home of the elusive cryptid known as Ogopogo. Reports of the event are contradictory, claiming that one man, a group of Indians, or campers from a local tourist ranch found the carcass beached on or "across from" Rattlesnake Island.[80]

Local author Frank Morgan Buckland was apparently the first to describe the creature, in his book *Ogopogo's Vigil: A History of Kelowna and the Okanagan,* published by the Okanagan Historical Society in 1948. According to Buckland, "One of the party who had gone to the lake edge for water was attracted by a strong smell of rotted fish. On investigation he found the badly decomposed body of a strange animal lying at the water's edge. The body was between five and six feet in length and would weigh about 400 pounds. It had a short, broad, flat tail and a head that stuck out from between shoulders without any sign of a neck. The nose was stubby, sticking out of a rounded head with no ears visible. The thick hide was sparsely covered with a silky hair four or five inches in length and of a bluish-grey colour while the teeth resembled those of a dog. It had two ivory-like tusks and claws resembling those of a great bird, on flipper-like arms; claws that showed no signs of wear or use, such as those of a cougar or other land animal."[81]

An unnamed "local amateur naturalist" examined the carcass, pronouncing it a manatee (genus *Trichechus*), despite the fact that those gentle aquatic herbivores (a) have no talons or "ivory-like tusks," and (b) live primarily in warm water, ranging from the Caribbean and Amazon Basin to West Africa. While equally at home in salt or fresh water, sometimes straying as far from their normal range as New York, the presence of manatee in Okanagan Lake remains unexplained. To reach it, the lost - and apparently mutated - stray must have entered the Columbia River at Astoria, Oregon, and proceeded from there on an epic journey winding west- and northward through Washington State, into the Okanagan River (a Columbia tributary), and on from there to Okanagan Lake above Kelowna. While it may not be impossible, the theory requires a massive leap of faith - and still fails to account for the claws and tusks described by Buckland.

Peter Costello rejected the manatee hypothesis in 1974, likewise ignoring the creature's supposed tusks and claws, to write: "It is my guess that the animal was actually an ogopogo, as the details of this mammal with flippers and a broad tail and dark colour are all that we would expect. But that the carcase was mangled so much that the long neck was already gone." In which case, we are forced to ask how Ogopogo's head was reattached - "stuck on its shoulders," as Costello says - after the neck was severed and removed.[82] The mystery remains unsolved.

Lake Champlain, New York: 1915

Traditions of a large cryptid in Lake Champlain - predictably dubbed "Champ" - reliably date from 1819. The first reported sighting occurred at Bulwagga Bay, near Port Henry, New York, and Champ apparently returned there for another visit nearly a century later. According to author Rosemary Guiley, "In 1915, several people saw the monster stranded in the shallows of Bulwagga Bay, struggling to get free, this report estimated Champ's length to be about 40 feet."[83]

Since Champ has yet to be captured or identified, we may assume that it escaped on that occasion, and thus does not rate globster status. And in fact, my research for *Globsters* produced an article from the *New York Times,* dated 19 April 1915, which confirmed Champ's successful escape on the previous Wednesday (14 April). According to that piece: "When first sighted through a field glass the 'serpent,' which one observer says was about forty feet long, was apparently stranded on the reef at the entrance to Bulwagga Bay, near the Crown Point fortifications. Presently he released himself and after a few wild plunges which lashed the water into foam, headed for the Vermont shore in great semicircular sweeps, finally sinking submarine fashion leaving a wake which was well defined on the glassy surface of the lake."[84]

Soldier Key, Florida: 1921

A globster overlooked by Heuvelmans in his compendium washed ashore on Soldier Key, five miles south of Key Biscayne, sometime in January 1921. According to a story in the *New York Times,* published on 13 February -

> Scientists and deep-water fishermen here are puzzled over the finding of a huge sea monster off Soldier's Key [*sic*], a few miles south of Miami, which they are unable to name. The body of the creature was found three weeks ago by Elmer E. Garretson of Huntington, L.I. [Long Island], N.Y., who today towed part of the skull to Miami. This fragment is fifteen feet long and seven feet wide and weighs three tons.
>
> Mr. Garretson said he did not know the length of the monster, but said he saw as much as eighty feet of it. Sharks were devouring the flesh when he first found it. Mr. Garretson asserted that the creature evidently came up out of the deep waters of the Gulf Stream to die. About six feet of the skull protruded from the water.
>
> Mr. Garretson will head a party which will try to bring the remaining portions ashore.[85]

No follow-up was forthcoming, and while Ben Roesch subsequently judged the carcass to be a whale's, with authors Loren Coleman and Patrick Huyghe likewise deeming it "whale-related," we are left to wonder why Miami's anonymous scientists could not identify it as such.[86] The largest known whales were described in the mid-eighteenth century, and should have been readily recognized from a giant skull.

Jehu Sands, India: 1921

In April 1921 a creature from the sea was beached at Jehu Sands, some ten to twelve miles from Bombay (now Mumbai), India. As Bernard Heuvelmans described it: "It was not a rotting corpse but a live animal which was still giving despairing cries forty-eight hours later. The British papers reported that it was 25 feet long and that its mouth, at least 3 feet deep, was lined with formidable teeth: it could easily have swallowed three men in one mouthful. Its skin, through which the ribs could be seen, was black. Its eyes, like an elephant's, rolled in their sockets in a way that seemed to bode no good. Its head was rather like a man's."[87]

From that, Heuvelmans concluded that it must have been a pilot whale (genus *Globicephala*), noting that one species - *G. indicus* - was found in the Indian Ocean. Today, that species is considered synonymous with *G. macrorhynchus,* the short-finned pilot whale, which enjoys worldwide distribution. That said, Ben Roesch reports that the record specimen of *G. macrorhynchus* measured eighteen feet long, while more generous estimates limit the species to twenty-one feet. Even at the greater length - or allowing an extra nineteen percent for an aberrant specimen, no pilot whale has a mouth three feet deep, capable of swallowing three men simultaneously. Nor, for that matter, do the teeth of *G. macrorhynchus* resemble those of the Jehu Sands creature, which seem more akin to those of a great white shark.[88]

Cape May, New Jersey: 1921

Cape May is our first locale to boast consecutive globsters, albeit thirty-four years apart. The second stranding occurred in November 1921 and was overlooked by the *New York Times*. Two years later, described for the first time by English author Frederick Albert Mitchell-Hedges in his book *Battles with Giant Fish*. He wrote: "This mammal, whose weight was estimated at over 15 tons, which - to give a comparison of size - is almost as large as five fully grown elephants, was visited by many scientists, who were unable to place it, and positively stated that nothing yet known to science could in any way compare with it. The photographs which were published in many newspapers showed that this modern leviathan somewhat resembled an elephant - in fact, it could be best described as a sea-elephant, but of huge proportions."[89]

The reference to baffled, unnamed scientists is naggingly familiar. Likewise, it seems the "many" photographs have not survived. By "sea-elephant," Mitchell-Hedges presumably means an elephant seal (genus *Mirounga*), whose southern species (*M. leonina*) boasts a record length of twenty-two feet six inches but - at least in theory - never strays from subantarctic waters. The northern elephant seal (*M. angustirostris*) is smaller, with a record length of fourteen feet, and has been documented only from the western shores of North America.[90]

Charles Fort took up the case in 1931, writing, "I investigated the story of the Cape May monster, wherever I got the idea that I could find out anything in particular. Somebody in Cape May wrote to me that the thing was a highly undesirable carcass of a whale, which had been towed out to sea. Somebody else wrote to me that it was a monster with a tusk twelve feet long, which he had seen. He said that, if I'd like to have it, he'd send me a photograph of the monster. After writing of having seen something with a tusk twelve feet long, he sent me a photograph of something with two tusks, each six feet long. But only one of the seeming tusks is clear in the picture, and it could be, not a tusk, but part of the jaw bone of a whale, propped up tusk-wise."[91]

No further data is available, but Fort's judgment seems sensible enough. Bernard Heuvelmans concurred in 1965, judging the animal to be a baleen whale, and Ben Roesch echoed that verdict in 1999.[92]

Margate, KwaZulu-Natal: 1922/24

This globster ranks among history's most perplexing, not only for its outlandish appearance, but because reported dates for its dramatic stranding span two full years. Stranger still, some accounts suggest that *two* different globsters were beached near the same location within days of one another, in completely unrelated incidents.

British authors John Michell and Robert Rickard (founder of *Fortean Times*) described the event most concisely in their book *Living Wonders* (1982). According to them:

> On the morning of 1 November 1922, Hugh Ballance looked out to sea
> from the beach at Margate, in Natal (South Africa), where he had recently
> bought a farm. A disturbance out to sea caught his attention, and through

A 1925 photo of "Trunko," uncovered by Dr. Karl Shuker in 2010.

his glasses he thought he saw "two whales fighting with some sea monster" which looked like a huge polar bear. According to a statement he made to a local newspaper, cited in the *Daily Mail* (27 December 1924), Ballance said: "This creature I observed to rear out of the water fully 20 feet and to strike repeatedly with what I took to be its tail at two whales, but with seemingly no effect." The battle continued for several hours watched by a growing crowd on the beach. Eventually the whales moved away leaving the strange giant floating without a sign of life. That night the carcass drifted ashore on a beach near the aptly named Tragedy Hill. The body was colossal and spread out upon beaching, as do all large sea creatures without their natural element to support their bulk. It was 47 feet long, 10 feet wide and 5 feet high. It had a 10-foot-long tail, matched at the other end by a curious trunk-like appendage. "Where the head should have been," said Ballance, "the creature had a sort of trunk 14 inches in diameter and about five foot long, the end being like the snout of a pig." But the most astonishing feature of the monster, which could be seen clearly from the beach during the previous day's battle, was its impressive fur or hair covering, "eight inches long and exactly like a polar bear's, and snow white." There was no sign of any wound or blood stains. For ten days it lay there on the beach, attracting sighteers and flies, until the stench became intolerable. A team of 32 oxen failed to move it far and abandoned it near the water's edge, from where the night tide wafted it back into unknown depths."[93]

Michell and Rickard theorized that the creature's trunk - which, inevitably, led it to be christened "Trunko" - might be the serpentine neck of some unknown species. They acknowledged that rotting shark and whale carcasses often present a "hairy" appearance, but balked at linking those fibrous bristles to Trunko's eight-inch pelt.[94] In any case, a creature freshly killed would not have decomposed enough within so short a time to make its hide unrecognizable.

At this point, we are moved to ask why a such a flamboyant monster, beached in such unique circumstances, would pass unreported from November 1922 until December 1924. The plot thickens with an anonymous article published in the August 1925 issue of Britain's *Wide World Magazine*. According to that piece, quoting one A.K. Jones of Johannesburg, Trunko was actually beached "around the middle of November last" - i.e., 1924 - "about three weeks" after witnesses reported "a terrific struggle at sea...between what they took to be a whale and some other animal they could not distinguish." Various observers took the creature for an octopus, a whale, or polar bear, despite its massive size. Jones said that Margate residents were "mystified" when the carcass vanished "about ten days later," adding that it beached a second time, three miles farther up the coast, "about the middle of December." Jones, on holiday at Margate, visited the second stranding sight and snapped two photos of the carcass that were published in *Wide World*.[95]

Some discrepancies in Jones's account - the date of Trunko's first stranding, the three-week lapse between its beaching and the battle at sea, the single whale observed in combat with Trunko - may all be explained by the photographer's belated arrival in Margate. Still, there appears to be no doubt that it was beached during November 1924. And yet ....

Charles Fort addressed the case in *Lo!*, writing:

> In the London *Daily Mail*, Dec. 27, 1924, appeared a story of an extraordinary carcass that was washed up, on the coast of Natal, Oct. 25, 1924. It was 47 feet long, and was covered with white hair, like a polar bear's -
>
> I won't go into this, because I consider it a worthless yarn. In accordance with my methods, considering this a foolish and worthless yarn, I sent out letters to South African newspapers, calling upon readers, who could, to investigate this story. Nobody answered. [96]

Authors Michell and Rickard, looking back at Fort was, may have been misled in dating the event by garbled newspaper reports, but something even stranger is afoot here. According to Bernard Heuvelmans, citing the March 1925 issue of *Wide World Magazine,* an entirely different globster was beached near Margate on 25 October 1924. As Heuvelmans explains:

> A South African farmer, Mr. W. White of Fascadale, informed the *Natal Mercury,* in Durban, that on October 25, 1924, he and his friend Frank Strachan had gone to examine a sea monster which, according to local Blacks was stranded on the beach at Baden-on-Sea, near Margate. Having seen the monster, they concluded that it was a "champion octopus."
>
> Their diagnosis is clearly wrong, for on the sketch attached to Mr. White's letter one recognizes a mutilated animal where there are still ten stumps clearly visible. It is thus a Decapod and, from its enormous size, probably *Architeuthis.*[97]

White's letter explained his examination of the carcass as follows:

I have taken careful measurements in the presence of Mr. Strachan and these are given on the sketch. You will notice the shortness of the tentacles. These would appear to have been bitten off by sharks. The object is lying on its back high and dry when the tied is out, and so far appears quite fresh (no smell). The body is very hard, so much so that Natives have been trying to cut it up, but their large knifes [*sic*] have had no effect. The whole body is white in colour.

I would say, from the thickness of the two front feelers, that the overall measurement would have been something like 50 ft had they not been eaten off. You will notice that the tentacle on the left is quite long, and I would say that the body to the tip would have measured 22 ft or more. The body is 9 ft in thickness.[98]

Perhaps we might suggest that White and Strachan were among the witnesses mentioned by A.K. Jones, who thought Trunko was an octopus. That suggestion founders, though, on White's sketch and description of the carcass he observed. Aside from being white, the two bodies share nothing in common. White's cephalopod is twenty-eight feet long and twenty-four feet wide, with ten partial appendages and no hair. Trunko is shaggy, forty-seven feet long and ten feet wide, with only two appendages - the supposed "trunk" and "tail." Quite clearly, based on the descriptions and on Jones's photographs, they cannot identical.

To confuse matters further, Heuvelmans included both globsters in the American edition of his sea-serpent volume, separated by 496 pages. He resolved any conflict by stating that Trunko was beached on 1 November 1922, then introduced further confusion in his appendix of globster strandings, writing that the carcass came ashore "before October" 1922. His source? The *Daily Mail* article from 27 December 1924![99]

A website maintained by the Margate Business Association tried to resolve the date discrepancy in October 2009, quoting "A.C. Jones of Johannesburg." Presumably, *Wide World Magazine* mistook his middle initial in 1925, since Jones confirmed taking the photographs of

Trunko and insisted that the carcass was beached in November 1924. He blamed two published sources - Thomas Victor Bulpin's *Natal and the Zulu Country* (1966) and an unspecified "pamphlet dealing with South Coast Resorts" - for mistaking the year. Jones apparently never saw Heuvelmans's book or *Living Wonders,* which opined that the *Daily Mail* article appeared two years after Trunko's stranding.[100]

Jones should know what he's talking about, and yet, the MBA's online summary of Trunko's case contains both revelations and discrepancies. After misspelling Trunko's name in its title, the article proceeds to say that the epic battle at sea occurred on 2 November, without mentioning a year. It further claims that all three combatants died, a switch from every other published version of the tale which has the whales emerge victorious. In addition to its trunk, the article contends that Trunko had "a 10 foot long lobster type tail all of which was covered in 8 inch snowy white hair." Next, it says that Trunko lay ashore for ten days, then vanished forever - a flat contradiction of Jones's statement to *Wide World Magazine* in 1925. Finally, the brief article says that the *Daily Mail* story appeared "over 2 years later," when local witnesses "still recalled the valiant battle against the lethal whales." Perhaps the unnamed author's style is to blame, presenting "the legend" of Trunko with snippets from Jones as a weak rejoinder.[101] In any case, the muddle remains - and the separate ten-armed carcass from October 1924 is once again ignored.

What was Trunko? The photos uncovered in 2009 (with another found by Karl Shuker in 2011) clearly destroy Charles Fort's suggestion that nothing was beached at Margate; ditto for Ben Roesch's dismissal of Trunko as "a newspaper hoax...on the grounds of zoological implausibility alone." They also clearly do not depict the "champion octopus" sketched by witnesses White and Strachan. Leaving that carcass aside, what remains? Dr. Heuvelmans could not suggest a candidate, tagging the case with yet another enigmatic question mark. Lance Bradshaw, writing on his Kryptid's Keep website, casts his vote for a decomposed whale, while admitting knowledge of "no creature, living or dead, that closely resembles this supposed animal." He closes by suggesting that we name Trunko *Maritomammuthus alba* ("white mammoth of the sea"), but grants that his proposal "is merely conjecture."[102]

So, finally, is any theory on Trunko, a true mystery of the sea.

Santa Cruz, California: 1925

Near the end of May 1925, five months after the *Daily Mail* first published news of Trunko, another carcass washed ashore on Moore's Beach (now Natural Bridges State Beach), at Santa Cruz, California. Its head was "longer than a man," according to Heuvelmans, while various published accounts placed its total length anywhere between thirty-five and fifty feet - pegged by Heuvelmans at thirty-six feet six inches. It seemed to have a slender neck, no less than twenty feet in length. While photos of the globster baffled all who viewed them, examination of its skull at San Francisco's California Academy of Sciences identified the creature as a specimen of Baird's beaked whale.[103]
Or so, at least, reads the official version of the tale.

One who disagreed with that judgment was zoologist E.L. Wallace, named by author Mark

The Santa Cruz carcass, 1925.

Chorvinsky as the only scientist to examine the Santa Cruz carcass *in situ.* From personal observation, Wallace declared that the beast had a bill but no teeth, that its body contained no bones large enough to be whale vertebrae, and that its three-foot tail was too short and weak for a bona fide deep-sea denizen. Instead, Wallace proposed a swamp-dwelling herbivore, perhaps some aberrant member of the normally carnivorous order *Plesiosauria,* entombed in Arctic ice since prehistoric times and slowly thawing on a slow drift southward.[104]

Local newspapers did their bit to muddy the waters. The *Santa Cruz Sentinel* offered an eyewitness account of a "gigantic fish" battling a dozen or more sea lions offshore, shortly before the carcass beached. The rival *Santa Cruz News* said that the globster's head was "bigger than a barrel," sporting eyes "larger than abalones." Other accounts added multiple pairs of legs to the corpse, with "ivory toenails." *Weird California* author Greg Bishop professes to see one such "elephant leg" in photos snapped of the carcass, seemingly invisible to others.[105]

Adult specimens of Baird's beaked whale do have teeth, but only two, located at the tip of the lower jaw. In theory, they might be dislodged from a wave-battered, decomposed specimen. Heuvelmans explained the globster's slender "neck" by speculating that: "Apparently the effect of decomposition and heavy seas had been to separate the body from the skin, which had rolled up on itself like a Swiss roll to form the illusion of a long neck joining it to the body, which had also been washed up a little distance away."[106]

Prah Sands, Cornwall: 1928

Two new globsters surfaced in June 1928, the first at Prah Sands (now Praa Sands, pronounced "pray"), a coastal village with a mile-long beach in Cornwall, England, midway between Penzance and Helston. The carcass came ashore on 7 June but was generally ignored until December 1933, causing Dr. Heuvelmans to misstate the date in his sea-serpent text (though he corrected it in the appendix).[107] The story finally drew widespread notice on 11 December 1933, when *The Times* of London ran a letter from witness E.J. Garmeson. It read:

I saw the dead body of a very curious animal washed up on shore at Praa Sands in the Spring of 1928. It had been killed by the storm which threw it up and the head had been torn off, but it must have been not unlike the Loch Ness "monster" as described by Commander Gould. Several feet of the snakelike neck and what remained of the body measured from neck to tail approximately 30 feet, while it was some 3 to 4 feet in diameter at the thickest part of the body barrel. There were four feet-like flippers for swimming and the tail tapered to a point. The colour was a dirty white with some traces of pink, the skin was coarse and covered with hair or bristles, while the bones, which were of considerable size, were more fish-like than mammal.

Unfortunately, the position was such that it was impossible to get a satisfactory photograph, and the situation was too out of the way for any proper investigation to be made.[108]
Another witness, one Major Hutchinson, compared the creature's neck to a giraffe's, with the head end jagged. Two flippers near the body's front end were eighteen to twenty-four inches long and rimmed with bristles, while coarse five-inch hair covered the carcass. Two more fins or flippers at the tail's end resembled those near the front. Responding to a report received from coastguardsmen, scientists at London's Natural History Museum declared the specimen a basking shark without a personal examination of the carcass. Forty years later, Dr. Heuvelmans concurred.[109]

Gulf of Fonseca: 1928

Nine days after the Prah Sands globster appeared, on 16 June 1928, the *New York Herald Tribune* reported another carcass discovered, this time on the Central American Gulf of Fonseca, bordering present-day El Salvador, Honduras and Nicaragua. While the article was titled "Reptile's Fossil Found," the article did not describe fossilized remains. In fact, according to the *Herald Tribune,* the creature was "marked with black and white stripes, was exceedingly corpulent and a horn protruded from its head. Its fangs were one and a half inches long."[110]

Confronted with that strange description, Dr. Heuvelmans could offer no potential real-life candidates - and neither, so far, has anyone else. The horn suggests a narwhal, but those cetaceans live year-round in Arctic waters and do not possess striped coats.[111] As for the *Herald Tribune*'s suggestion that the creature was a "prehistoric reptile," several members of the order *Ceratopsia* had horns, but all were certified landlubbers. Another mystery remains unsolved.

Agulhas Bank: 1930

On 31 January 1930, the research vessel *Dana* was trawling for deepwater specimens west of the Agulhas Bank, a broad, shallow part of southern Africa's continental shelf extending 160 miles south of Capt Agulhas, between Quoin Point and Cape Infante. From a depth of 900 feet, the net produced a leptocephalus - eel larva - of an unknown species. It measured six feet one inch long. The specimen was bottled and preserved for display at Copenhagen University's Zoological Museum.[112]

Each species of eel has a distinct leptocephalus ("slim head"), a flat and frequently transparent larval form which develops to adulthood in the manner of a tadpole becoming a frog. The *Dana*'s crew was understandably excited by their find, since the leptocephalus of a common or European eel (*Anguilla anguilla*) measures three inches long and produces adults with a record length of four feet six inches. If the *Dana*'s leptocephalus grew in the same proportions, its adult form could be 108 feet long - a veritable sea-serpent indeed.[113]

There was, of course, a catch. Various species of eel develop from their leptocephali in radically different proportions, some barely gaining any length at all, while others - in the style of some amphibians - may actually be *shorter* than their larval forms. It is irrelevant in any case, since University of Miami ichthyologist Dr. David Smith reported in 1970 that the *Dana* leptocephalus belongs to a spiny eel (family *Notacanthidae*), which is not a true eel at all, but rather a bottom-dwelling fish whose adult specimens range from eight inches long to a record length of three feet nine inches.[114]

Glacier Island, Alaska: 1930

Glacier Island lies in Prince William Sound, thirty-five miles west-southwest of Valdez, Alaska. It came to world attention for the first time on 26 November 1930, when the *New York Times* published an article under the headline "Ice Bares Strange Animal; Alaskans Suggest Prehistoric Origin - Museum Here Investigating." According to that piece, the frozen beast was forty-two feet long, including a six-foot-long head, a twenty-foot body, and a sixteen-foot tail. The *Times* advised its readers that "[t]he theory has been advanced that the carcass is that of a prehistoric animal or reptile that has been preserved in the upper reaches of the Columbia glacier."[115]

Other Gotham papers quickly followed up on that report. On 28 November the *New York Sun* declared, "Monster in Ice Has Long Snout." Another *Times* competitor, the *Evening World,* printed a story on the same day, headlined: "Confirm Finding of Pre-Historic Monster in Ice." Intrigued by the discovery, W.J. McDonald, supervisor of Alaska's Chugach National Forest, led six companions to find and examine the carcass. MacDonald's measurements described a creature with a head resembling an elephant's, fifty-nine inches long, complete with a snout emerging from its forehead, thirty-nine inches long and eleven inches in diameter at its midsection, with a twenty-nine-inch circumference. The mostly-skeletal remains measured thirty-eight inches across at the widest point. Overall, the carcass was twenty-four feet long, with a fourteen-foot tail beginning at the rib cage. Years later, Charles Fort claimed that the carcass had "considerable flesh" remaining, but McDonald's report states that "only a small portion of the body had flesh on it," that remnant resembling meat from a horse. According to McDonald, the creature was noteworthy for its "long tail and tapering head, like a dinosaur."[116]

Typically, newspaper coverage of the discovery soon petered out. One final item from the *Ogden* (Utah) *Standard-Examiner,* published on 11 January 1931, ventured to describe the unknown creature's last moments: "Clambering clumsily from floe to floe, stopping now and then to assure his footing, now and then to gaze about at this north country in which he was a stranger, a giant creature, covered with glossy fur, made his way toward Glacier Island. He blinked his eyes - they were larger than a man's head - and rested. Then he looked forward again and saw a large table of ice. He made for it. His head measured six feet from nose to neck, and from mouth to tail tip he measured over fifty feet. The surface which would hold him in repose, without sinking, would have to be a large one. He reached the glassy raft which his eyes had detected - and luck was with him. It was large enough to lay down and he went to sleep."[117]

If the creature was a reptile, the *Standard-Examiner* opined, "it must have been encased before the general migration began. This is inevitable, since a single member of a species could not exist long alone." But *was* it a reptile? The paper's own mention of "glossy fur" argued otherwise, though previous reports offered no such detail. A coat of seeming "hair" might be explained by the decomposition of a basking shark, but that diagnosis renders the massive rib cage anomalous. Bernard Heuvelmans had no suggestions to offer in 1965, nor did Mark Chorvinsky, two decades later.[118] It finally remained for Dr. Karl Shuker to solve the mystery in September 2010.

Specifically, Shuker uncovered long-lost documents detailing the fate of the Glacier Island skeleton - purchased for $600 by the owner of a taxi company in January 1931 and displayed as a traveling carnival exhibit throughout North America before it was finally lodged for good at the Smithsonian Institution's National Museum of Natural History in Washington, D.C. There it remains today, identified as the remains of a minke whale or lesser rorqual (genus *Balaenoptera*).[119]

Newfoundland, Canada: 1932

Dr. Heuvelmans reported out next case in 1965, without citing a source or providing much detail. His brief account tells us that "[i]n May 1932, no doubt because the water had been poisoned, the shores of Newfoundland were covered with millions of dead fish. Among them was a huge beast with a pointed snout and sharp teeth and looking very like a snake." After due consideration, Heuvelmans tagged it with another nagging question mark.[120] At press time for *Globsters,* I was sadly unable to discover any new information on that most intriguing case.

Querqueville, France: 1934

More detailed coverage was offered for the globster found by fishermen at Querqueville, a commune in the Manche department of northwestern France, on 28 February 1934. As described by Heuvelmans, "Its head, which they thought looked like a camel's, was on the end of a long thin neck. It had two big flippers at the front of an elongated body, a fin on its back and a long tapering tail. Its skin seemed to be covered with a fleece of short stiff whitish hair. It was about 20 feet long, of which 3 feet were neck, and it was 5 feet in diameter at the thickest point. Some 60 yards away there lay a heap of entrails, which according to *La Croix* of Paris consisted of the beast's lungs, peritoneum and kidneys."[121]

The Querqueville carcass.

As photos of the carcass circulated, various experts offered opinions as to the creature's identity. Dr. Burgess Barnett, curator of reptiles at the London Zoo, claimed it was a small whale, species unknown.

At London's Natural History Museum, zoology curator Dr. W.T. Calman cast his vote for a basking shark, while colleague Martin Hinton suggested a giant seal. Dr. Georges Petit from the Paris Museum was the first scientist to examine the carcass *in situ*, quickly naming the mutilated beast as a basking shark. Thus was another mystery resolved.[122]

Rottnest Island, West Australia: 1934

The date of this peculiar stranding is uncertain, since the incident avoided public notice for the better part of thirty years. It only came to light in March 1962, thanks to publicity belatedly surrounding the discovery of a globster beached on Tasmania's shore in August 1960 (see below). By then, witness R.H. Timberly, a resident of Perth, could only say that his discovery occurred sometime in 1934.[123]

Timberly and his father made their find on Rottnest Island, located eleven miles offshore from Fremantle. The blob of flesh they found on shore was eighteen feet long and "roughly stingray shaped, with a long tail and vaguely formed flippers." Its skin was covered by "strange wool," while its "creamy flesh had the consistency of tough tripe." Photos snapped by Timberly and saved for three decades appeared to show "frontal bones and a toothless mouth." He also kept samples of the creature's flesh, which apparently were never analyzed. Based on the photos and description, author Malcolm Smith concludes that the creature was a giant ray, species unknown.[124]

Henry Island, Washington: 1934

Nine months after the Querqueville stranding, another globster came ashore on Henry Island, in the San Juan Islands. While frequently described as part of British Columbia, Henry Island is actually part of Washington State's San Juan County - a collection of islands in the Salish Sea, normally counted as part of the larger San Juan Archipelago.[125]

Fisherman Hugo Sandstrom found the carcass in November 1934. The first report of its discovery, published on 22 November, quoted observations by Dr. Neal Carter, director of the dominion fisheries experimental station at nearby Prince Rupert. According to Dr. Carter, "The creature was about 30 feet long. Red flesh indicated it was some sort of a warm-blooded marine mammal. It had a head shaped somewhat like that of a horse, and a tough, rough skin. The upper part of the skin bore hair and the lower part quills, like spines. The only bone of importance was the backbone."[126]

In Carter's estimation, the beast had died about two months before it washed ashore. As to what it was, he and his colleagues were uncertain. Their best suggestion was that "[s] cientific classifications indicated that the creature might have resembled the ichthyosaurus or the proteosaurus, an order of extinct marine creatures known to have existed in the Mesozoic age, when animal life on earth supposedly progressed from fishes to reptiles." That notion contradicted the mammalian diagnosis, but it satisfied Professor Trevor Kincaid, a marine biologist at Seattle's University of Washington. He told reporters, "If Dr. Carter, himself a well-known zoologist, has difficulty in determining what type of animal it was, there must be something truly significant about it."[127]

Alas, such did not prove to be the case. The globster's head, together with some of its flesh and vertebrae, were shipped off to Nanaimo, British Columbia, for study at the Pacific Biological Station of the Department of Fisheries, where Dr. W.A. Clemens swiftly identified them as the remains of a basking shark.[128]

Narooma, New South Wales: 1935

Narooma is located on the far south coast of New South Wales, historically a fishing port and cannery. Tuna and salmon are its stock in trade, but a very different creature came ashore within two miles of town on 15 April 1935, first seen by brothers Keith and Joseph Thompson. As described in the *Sydney Morning Herald,*

The animal was about eight feet long. It had a long, tapering, bony head. The lower jaw was studded with about 48 teet, which were separated by half-inch gaps. Most of the teeth were missing from the top jaw. The eyes were set just behind the gape of the large mouth. There were two fins just behind the head, a large dorsal fin, and two horizontal fins at the end of the tail. The whole animal was covered with a smooth, leathery skin.[129]

Australian naturalist David Stead opined that it was "probably" a dolphin, though it seems that he did not examine the carcass himself. Six decades later, author Malcolm Smith concurred, while suggesting that the species "might have been rare, or even new to science."[130] Since the remains were not preserved, it is impossible to say.

Twofold Bay, New South Wales, 1935

Nineteen days later, and a few miles farther south, another carcass came ashore at Twofold Bay, famous for orcas that supposedly helped whalers of the nineteenth century in their pursuit of larger cetaceans. The skeleton of one such helpful predator, dubbed "Old Tom," may be seen at at a museum in Eden, on the bay's north shore.[131] Concerning the globster that beached in early May 1935, the *Sydney Morning Herald* said:

> Mr. Michael Fourter, of Shadrach's Creek, near Eden, reports that some days ago he saw on the Boyd Town Beach, on the western side of Twofold Bay, the carcass of a marine animal, unlike any he had previously seen or heard of. It was, he said, about 8 feet long, and the head resembled that of a horse. The neck was arched. On the body were five fins, which were split so as to appear to consist of seven segments. On the inner side of the fins was a very hard bristle, white in colour. The flesh that was left was much like that of a fish. At the tail were two long bones, protruding in the shape of a cow's horns, and the ribs were about the size of a calf's ribs. The description resembles, in some respects, that of the strange beast found recently on a beach near Narooma, but whether the two animals were of the same species is not known.[132]

Malcolm Smith rightly notes that the two globsters, in fact, bore little or no resemblance to one another. He proposes that the second creature may have been a shark, but his suggestion that its gill slits were mistaken for calf-sized ribs is unconvincing - and still leaves the two bones at the tail completely unexplained.[133]

Camp Fircom, British Columbia: 1936

Thirteen years before his revelations on Trunko, Karl Shuker caused another cryptozoological sensation by unearthing a forgotten picture-postcard of a globster stranded at Camp Fircom on 4 October 1936. Run by the United Church of Canada, Camp Fircom is located on Gambier Island in Howe Sound, six miles north of West Vancouver's Horseshoe Bay Community. The photograph, first published in Shuker's book *The Unexplained*, depicts big-headed skeletal remains that bear a strong resemblance to another carcass found nine months later at Naden Harbour, British Columbia.[134]

While Shuker did not discuss the carcass in his text, the photo generated speculation that it might represent a juvenile specimen of *Cadborosaurus* - or "Caddy" - a marine cryptid seen frequently in Pacific Northwest waters, named by a Canadian newspaper editor in 1933 for the frequency of its appearances in and around Vancouver Island's Cadboro Bay.[135]

Camp Fircom's globster - a baby *Cadborosaurus?*

Vertebrate paleontologist Darren Naish took a contrary view in 1997, examining the print in Dr. Shuker's book and concluding that the photo was a hoax. In his view, it depicts "the stem of a large plant, most probably a kelp," artfully arranged to mimic a spinal column. He could not determine what the globster's skull and fins might be, and after hearing both his parents reject the kelp thesis, concluded that "[p]erhaps I am wrong." Dr. Shuker sent Naish a second, still unpublished post-card photo of the globster, but Naish remained committed to his thesis. In closing, he wrote, "I am not convinced, but then I have not seen the original, whereas Dr. Shuker has."[136]

Naden Harbour, British Columbia: 1937

The kelp hypothesis cannot explain another supposed Caddy carcass, retrieved from the stomach of a sperm whale in July 1937. The find was made by workers at the Consolidated Whaling Corporation's Naden Harbour station, on the northern coast of Graham Island, in the Queen Charlotte Islands, separated from British Columbia's mainland by Hecate Strait. The whale in question had been slain off Langara Island - also in the Queen Charlottes - and towed to Naden Harbour through Parry Passage, the trip spanning roughly ten days.[137]

What the flensers found upon gutting the whale was a creature beyond their experience, so strange that it prompted G.V. Boorman, the station's first-aid officer, to snap two photographs depicting what he called "The remains of a Sperm Whale's Lunch, a creature of reptilian appearance 10 ft. 6 in. in length with animal like vertebrae and a tail similar to that of a horse. The head bears resemblance to that of a large dog with features of a horse and the turn down nose of a camel." Another witness, station manager F.S. Husband, also photographed the carcass, which appeared to have been swallowed whole.[138]

Enter Francis Kernode - then serving as director of Victoria's Provincial Museum - who, in 1934, had declared the Henry Island basking shark to be a specimen of Steller's sea cow (*Hydrodamalis gigas*), a large sirenian presumed extinct since 1768. Controversial at best, accused by museum historian Peter Corley-Smith of burning corresondence that criticized his rash judgments, Kernode told the Victoria *Province* on 23 July that "there was little doubt that the portions of a backbone, the piece of baleen and the portion of skin forwarded to the museum were pieces of a baleen whale, which he believed was of premature birth." Dr. Ian McTaggart-Cowan, curator of vertebrate animals at the provincial museum - whose Ph.D. in zoology presumably trumped Kernode's background as a taxidermist - strongly disagreed, declaring that the Naden Harbour carcass represented nothing known to science. Subsequent enquiries failed to locate any trace of the peculiar creature. Its identity remains a mystery.[139]

Provincetown, Massachusetts: 1937/39
Our next case is another mired in confusion, with even the date uncertain. In the appendix to his treatise on sea-serpents, Heuvelmans claims that a carcass was beached sometime during 1939, at or near Provincetown, Massachusetts. He ventures no guess as to its identity, makes no mention of the event in his text, and offers no source for the report.[140]

From that anemic beginning, I pursued the case online, discovering an article written by ichthyologists William Bigelow and William Schroeder in 1948. That document describes a basking shark skeleton beached at Provincetown, examined by both authors, which measured twenty-five feet long. Unfortunately, through some careless error, separate paragraphs claim that the stranding occurred in January 1937 and in January 1939. Stranger still, a footnote for the former date directs readers to an article written by Schroeder for the now-defunct journal *New England Naturalist* in 1939.[141] Whichever date is correct, in fact, it seems safe to mark this mystery at least conditionally "solved."

Kitsilano Beach, British Columbia: 1941
Kitsilano Beach, on Burrard Inlet, is one of Vancouver's most popular beaches in warm summer months. In early March of 1941, however, it played host to a surprise guest dubbed "Sarah the Sea Hag" by local newspapers and touted as yet another possible *Cadborosaurus* specimen. According to the Vancouver *Province* of 5 March, "She had a large horse-like head with flaring nostrils and eye sockets; a tapering snake-like body 12 feet long; and traces of coarse hair on the skin."[142]

Ichthyologists W.A. Clemens and Ian McTaggart-Cowan examined the globster, tagging it a selachian. "We're not sure if it is a basking shark," Clemens told the *Province,* "but there is no doubt it is of the shark family." G.V. Boorman, from the Naden Harbour whaling station, strenuously disagreed. Based on his personal familiarity with sharks and his examination of stomach contents from 4,000 whales, he told the newspaper, "If that's a shark, I'll eat my uniform. I've seen the skeletons of scores of various sharks and they had no resemblance to these remains." In fact, Boorman declared, the Naden Harbour carcass of 1937 was "the twin sister of odiferous Sarah."[143] The remains were not preserved, and there the matter rests.

Deepdale Holm, Orkney Islands: 1941

A week before Christmas 1941, a witness remembered only as "Mr. Anderson" found a large globster beached at Deepdale Holm, 2.5 miles from the village of Holm (called "St. Mary's" in some accounts), on Mainland, largest of the Orkney Islands. Inspector Cheyne of the Kirkwall Police examined the body and described it as follows:

> From head to tail it is 24 ft. 8 inches long but it must have been larger than that because part of the head is torn away and some of the tail appears to be broken off. The hair on the body resembles coconut fibre in texture and colour. The monster has a head like a cow, only flatter. The eye sockets are three inches in diameter and very deep. The neck, which is triangular, is 10 ft. long and 2 ft. round, and at the base of the neck

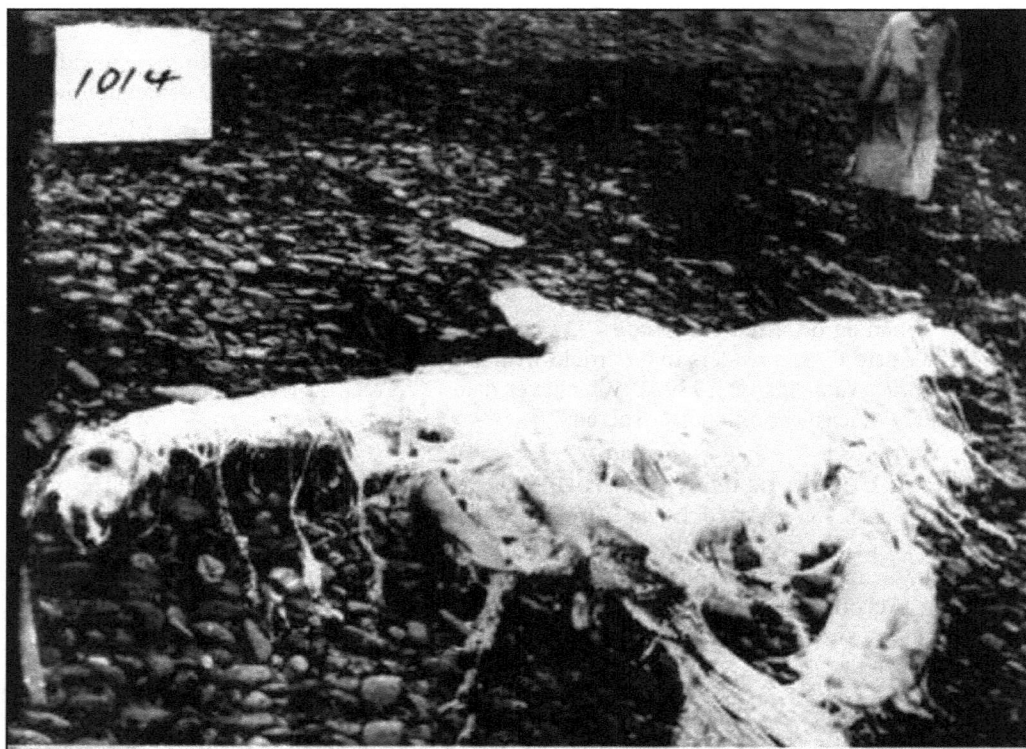

The globster stranded at Deepdale Holm.

there is bone, shaped like a horse collar and about 4 ft. thick. Across the back is a fin which is 10 inches thick at the base and tapers. This fin is two feet six inches high. Four flappers with bone structures like hands, each 3 ft. 8 inches long were apparently the monster's means of propulsion. The tail is pointed. There is at least a ton of flesh still on the body.[144]

Witness Anderson's wife contacted J.G. Marwick, provost of Stromness and a respected naturalist, describing the carcass as "a funny sort o' shark with a neck at both ends."[145] Marwick himself published a description of the creature, which differed from Inspector Cheyne's, in his "Nature Notes" column for the *Orcadian* newspaper. Therein, he wrote:

The neck from the base of skull to where it joined the body was 3½ feet roughly and 6 inches broad. A small hump was visible along the back, less than a foot high, then came the large hump or dorsal fin, which was quite intact and one of the more prominent features. This hump was 2½ feet high - and on its upper edge were several thick hairs much worn and broken by rubbing on the beach; the hump was rounded on top and curved towards the head and tail of the animal - being 2 feet 6 inches wide at the base. Proceeding towards the tail were a whole series of small projections plainly visible for several feet along the back, but not at all in the nature of the large hump already mentioned. Then, 7½ feet from the centre of the large hump was a third hump, not quite so prominent as the smaller one [*sic*] in front, but a distinct hump nevertheless. From this to the tail were more of the small projections, but at the extreme end of the tail the flesh was all gone, leaving the backbone in sight, several sections of which were broken off and what had been the utmost end of the tail was a piece of gristle - as I saw it the total length of the creature was 25 feet.[146]

In his next column, one week later, Marwick christened the beast *Scapasaurus*, after its close proximity to Scapa Flow, writing: "I judge it to be one of a species considered extinct many ages ago - a marine saurian in fact - or marine reptile. The very nearest I can come to it is the order of extinct reptiles called 'Ichthopterygia.'"[147] Marwick misspelled *Ichthyopterygia*, more commonly known as *Ichthyosauria* - and he was wrong, in any case. On the very day he named the *Scapasaurus*, spokesmen for Edinburgh's Royal Scottish Museum identified the creature as a basking shark.[148]

Deepdale Holm, Orkney Islands: 1942

The ink was barely dry on the report dismissing *Scapasaurus* when another carcass beached on Hunda, an uninhabited island in the Orkneys, lately linked to nearby Burray by a causeway built in 1941. Three observers - Campbell Brodie, Andrew Laughton and James MacDonald - viewed the carcass on 5 February, and Brodie described it in a letter to the *Orcadian*, published on 12 February. It was, he wrote, a "huge elongated yellowish-coloured creature" wedged between seaside boulders. Only the skull and spine remained, with several cartilaginous appendages, some "greyish-black tough skin with a hairy appearance," and two four-inch "antennae" on the skull. The carcass measured twenty-eight feet overall, with a two-foot-long skull and the largest fin three feet in length.[149]

Alas, the beast was not a second *Scapasaurus* - or, to be precise, it was. By 19 February it had been identified conclusively in the *Orcadian* as yet another basking shark.[150]

Gourock, Scotland: 1942

Scotland's third globster within half a year surfaced at Gourock, on the upper Firth of Clyde, in summer 1942. The date of its discovery is unrecorded, but council officer Charles Rankin and another (unnamed) man found the carcass while seeking the source of an odious stench on the shoreline. Wartime restrictions on seaside photography prevented any snapshots of the body being taken - and, in fact, Royal Navy officers ordered Rankin to remove the carcass and destroy it, without preserving any part for later study.[151] Still, we have his subsequent description of the beast.

> It was approximately 27-28 feet in length and 5-6 feet in depth at the broadest parts. As it lay on its side, the body appeared to be oval in section but the angle of the flippers in relation to the body suggested that the body section had been round in life. If so, this would reduce the depth dimension to some extent. With head and neck, the body, and the tail were approximately equal in length, the neck and tail tapering gradually away from the body. There were no fins. The head was comparatively small, of a shape rather like that of a seal, but the snout was much sharper and the top of the head flatter. The jaws came together one over the other and there appeared to be a bump over the eyes - say prominent eyebrows. There were large pointed teeth in each jaw. The eyes were comparatively large, rather like those of a seal but more to the saide of the head.

> The tail was rectangular in shape as it lay - it appeared to have been vertical in life. Showing through the thin skin there were parallel rows of "bones" which had a gristly, flossy, opaque appearance. I had the impression that these "bones" had opened out fan-wise under the thin membrane to form a very effective tail. The tail appeared to be equal in size above and below centre line.

> At the front of the body there was a pair of "L"-shaped flippers and at the back a similar pair, shorter, but broader. Each terminated in a "bony" structure similar to the tail and no doubt was also capable of being opened out in the same way.

> The body had over it at fairly close intervals, pointing backwards, hard, bristly "hairs." These were set closer together towards the tail and at the back edge of the flippers. I pulled out one of these bristles from a flipper. It was about 6 inches long and tapered and pointed at each end like a steel knitting needle and rather the thickness of a needle of that size, but slightly more flexible. I kept the bristle in the drawer of my office desk and some time later found that it had dried up in the shape of a coiled spring.

The skin of the animal was smooth and when cut was found to be comparatively thin but tough. There appeared to be no bones other than a spinal column. The flesh was uniformly deep pink in colour, was blubbery and difficult to cut or chop. It did not bleed, and it behaved like a thick table jelly under pressure. In what I took to be the stomach of the animal was found a small piece of knitted woollen material as from a cardigan and, stranger still, a small corner of what had been a woven cotton tablecloth - complete with tassels.[152]

One bristle aside, the carcass was hacked into pieces and buried without being viewed by any qualified zoologist.[153] The creature's prominent teeth rule out a basking shark, along with any other filter-feeding species, while no specimen of carnivorous shark has yet been documented at the size attributed to the Gourock globster. Likewise, the lack of bones apparently rules out a bony fish, reptile, or mammal. What, then, was the creature?

We shall never know.

Machrihanish, Scotland: 1944

Scottish wartime carcass beached at Machrihanish, on the Atlantic coast of Argyll and Bute, on 2 October 1944. The *New York Times* described it as a "furry beast 20 feet long," with huge eyes and feet. Dr. A.C. Stephens of the Royal Scottish Museum reportedly identified it as a basking shark, but if so, that verdict eluded Bernard Heuvelmans twenty years later. His list of stranded carcasses leaves the creature unidentified, tagged with another of his haunting question marks.[154]

Kenai Peninsula, Alaska: 1946

On 25 October 1946, Michigan's *Traverse City Record-Eagle* ran a story from the United Press news wire, headlined "Prehistoric Monster Found in Alaska." According to that article -

Anthropologists from the University of Alaska at Fairbanks were enroute here today to examine the body of a huge, lizard-like creature identified tentatively as a prehistoric tyrannosaurus or gorgonosaurus.

It was believed to have been preserved in a glacier until washed ashore here Wednesday [i.e., 23 October]. Although positive identification by experts has not yet been made, Fairbanks physicians studying anthropology texts said the "creature" was "definitely prehistoric" and may belong to one of two species."

The creature measured nearly nineteen feet from tip to tail. It's head measured 2 feet by 2½ feet and its mouth featured a row of teeth 18 inches long.

The animal had large hind legs and a heavy thigh bone which measured, approximately 4 feet from the hip to the first joint. The forelegs were short and heavy.

Leathery skin on the head and neck was covered with bristly hair and flesh almost completely covered the head, shoulders and hips. The backbone had broken through the animal's side and there was some evidence of decomposition.[155]

That description notwithstanding, Bernard Heuvelmans subsequently identified the specimen as a "Pacific grampus," citing as his source articles published in the October 1949 issue of *Natural History,* and in London's *Daily News* on 1 April 1950. He did not discuss the case beyond dismissing it in an appendix to his work on sea-serpents, and thereby left his readers with a puzzle.[156]

The "grampus" label may apply to either of two separate marine mammals. One, Risso's dolphin (*Grampus griseus*), is the only member of the genus *Grampus.* Adults average ten feet in length, with a record length slightly exceeding fourteen feet, and none possess eighteen-inch teeth. Earth's other "grampus" is the killer whale (*Orcinus orca*), though the former name is rarely used today. Killer whales commonly exceed twenty feet, with a record length of thirty-two feet, but their teeth, while large, average only three inches long.[157] Even if we accept the description of eighteen-inch teeth in a two-foot long skull as a typographical error, the matter of "large hind legs" with four-foot "heavy thigh bones" would seem to eliminate any known cetacean from consideration in this case.

Barkley Sound, British Columbia: 1947

Barkley (or Barclay) Sound lies on the southwestern coast of Vancouver Island, between Ucluelet and Bamfield. In the latter part of 1947 it played host to a globster whose story remains confused today, despite final agreement on the creature's identity.

According to Bernard Heuvelmans, the carcass came ashore at Effingham - actually Effingham Inlet - on 2 November. That date is almost certainly erroneous, since Heuvelmans places the stranding "soon after" a sea-serpent sighting off Ucluelet "in November," and both

Carcass beached at Barkley Sound, British Columbia.

sources cited for coverage of the event are dated 8 December 1947. Fisherman Henry Schwarz found the globster, recruiting three others to help him remove it in sections, transporting the whole to Port Alberni. Heuvelmans describes the carcass as including a toothless skull fourteen inches long and 8.5 inches wide, with a spine "at least 40 feet long," consisting of 145 cylindrical vertebrae.[158]

Two ichthyologists from the

Nanaimo Biological Station, Dr. R.E. Foerster and Dr. A.L. Tester, examined two of the beast's vertebrae and told the *Vancouver Daily Press*,

> "We believe, when we have made an examination of the remainder of the carcass which is being brought to Alberni, we will be able to say definitely that it is a huge ribbon fish. We were able to discern evidence of a long fin, which extends the length of the ribbon fish. One of the discoverers told us that when he first saw the carcass, it was lying on the beach and that a long dorsal fin extended from the head to the tail. This is characteristic of the ribbon fish." That said, the experts declared that "some parts are of a type similar to a shark skeleton, but the head is too small and the vertebrae too numerous for it to be a shark."[159]

In fact, Heuvelmans says that Foerster and Tester subsequently identified the creature as a basking shark - which, if true, not only cotradicted their initial judgment but ranked it among the largest specimens of *Cetorhinus maximus* on record. Paul LeBlond and Edward Bousfield concur in the final diagnosis, while adding variations to the story.

In their account, the carcass was forty-five feet long, with 150 vertebrae, and it remained for Dr. Clifford Carl at British Columbia's Provincial Museum to overrule the ribbon fish I.D. They also say the globster's skull was twelve inches wide, citing an article from the *Vancouver Sun,* published on 8 December 1947. Their account places the stranding at Vernon Bay, rather than Effingham Inlet.[160]

Dunk Island, Australia: 1948

Dunk Island lies 2.5 miles offshore from Mission Beach, Queensland, in the Coral Sea. It is part of Australia's Family Islands National Park, and also falls within the larger Great Barrier Reef World Heritage Area. Its total area is slightly less than four square miles, with its highest point - Mount Kootaloo - standing 880 feet above sea level.[161]

Sometime in 1948, according to the website Cryptopedia, a globster washed ashore on Dunk Island. We know nothing of the case today, except that "the carcass is said to have been tough and furry and resistant to fire." No source was cited for this observation, and none could be located by press time for the work in hand.[162]

Ataka, Egypt: 1950

In early January 1950 a three-day storm lashed the Gulf of Suez. When it cleared, residents of Ataka, Egypt, found a large carcass washed ashore near their town. It was "whale sized" and possessed an apparent blow-hole, but two huge tusks protruded from its head. Various published accounts indicate that a "team of scientists" arrived to examine the creature but left it without making a positive identification. Nonetheless, all extant reports seem to agree that the beast was some kind of whale, and that the "tusks" were its jawbones denuded of flesh. Dr. Heuvelmans confidently identified the Ataka creature as a specimen of Bryde's whale (*Balaenoptera brydei*). Thus far, no one has presumed to contradict him.[163]

Devils Lake, Oregon: 1950

Devils Lake is a modest body of water: three miles long, one-third of a mile wide, and twenty-one feet deep at its deepest point. It lies in Lincoln County, Oregon, linked to the Pacific Ocean by the 120-foot-long D River. Prior to 1989, *Guinness World Records* listed the D as the world's shortest river, but it still managed to deliver a surprise on 4 March 1950.

That morning, residents of nearby Delake - a town that no longer exists - discovered a large carcass stranded on the lake's shore. Observers described the body as twenty-two feet long, covered with hair, weighing an estimated 1,000 to 3,000 pounds. One report added feathers to the creature's hair; another claimed that it "had the body of a cow" and "approximately nine tails." No one thought to snap a photograph, apparently, but locals nicknamed the beast "Old Hairy."[164]

Experts differed in their assessment of Old Hairy's identity. An unnamed biologist from the Oregon Fish and Wildlife Commission branded the object a lump of whale blubber. Professor Fred Kohlruss from the University of Portland disagreed. "It's an elasmobranch," he said. "It's a sea inhabitant whose bones remain in the cartillage stage." That narrowed the field to sharks, rays and skates, without suggesting a species. Finally, Dr. E.W. Gudger, speaking for the American Museum of Natural History, declared the beast to be a whale shark.[165] While plausible enough in theory, no part of the carcass was preserved for confirmation, nor does any version of the tale explain what finally became of Old Hairy's corpse.

Hastings, New Zealand: 1951

This settlement lies on the shore of Hawke Bay, on the southwestern coast of New Zealand's North Island. On 11 January 1951, two newspapers in Wellington reported the discovery of thirty-foot skeletal remains stranded near Hastings, on the Waimarama coast. The skull measured three feet six inches wide and sported a three-foot-long "tusk," subsequently identified as a beaked snout.

Amidst speculation that some deep-sea "disturbance" had killed the creature and cast it ashore, a Mrs. F.O. Bryce announced that she had seen a fifteen-foot beaked whale in Polorus Sound, one of the Marlborough Sounds at the northern end of New Zealand's South Island. Had a specimen twice that size found its way farther north before dying, decomposing, and drifting ashore? The mystery remains unsolved.[166]

Hendaye, France: 1951

Hendaye is the most south-westerly town in France, a popular tourist resort on the "Côte Basque," overlooking the Atlantic Ocean from the Pyrénées-Atlantiques department. In February 1951 a rotting carcass came ashore, described as sixteen feet in length and looking "just like a prehistoric animal." Witness Léon Duourau told Bernard Heuvelmans that the carcass had "a tortoise's head with two cartilaginous antennae, a long plesiosaur's neck, a brown body with large oval scales in places, and four short webbed feet like a

A sketch of the Henday carcass, 1951.

seal's or a turtle's." A photograph of the remains convinced Heuvelmans that the animal had been a basking shark, its "antennae" nothing more than rostral cartilage supports for the shark's pointed snout. As for the scales - which sharks do not possess - they may have been loose bits of decomposing skin.[167]

Shuyak Island, Alaska: 1951
Five months after the discovery at Hendaye, on 21 July, the *Kodiak Mirror* reported another carcass washed ashore on Shuyak Island, in southern Alaska's Kodiak Archipelago. No description of the creature is available today, but Heuvelmans dismissed it as an unidentified cetacean.[168]

Old Bar, New South Wales: 1952
Details are sparse concerning a carcass beached near this town, at the mouth of the Manning River, sometime in 1952. Author Malcolm Smith, citing an undated clipping from the *Sydney Morning Herald,* reports that it was fifteen feet long and eight feet eight inches in circumference. It had "a head like a calf, an elongated duck-shaped bill, two small flaps on either side, and a fan-shaped tail." Local residents thought it might be a dugong (*Dugong dugon*), but the curator of mammals at the Australian Museum, a Dr. Troughton, reminded them that dugong's rarely exceed nine feet in length. Although he never viewed the carcass personally, Troughton ventured the opinion that it might be "some unusual form of whale."[169]

Whitburn, England: 1953
Whitburn is a village in South Tyneside, formerly County Durham, located on the coast of North East England. Dr. Heuvelmans reports that an oarfish beached there in February 1953, but he offers no description of the carcass, nor any source for his report.[170]

Canvey Island, England: 1953

Canvey Island, in the Thames estuary, is a reclaimed island separated from the coast of south Essex by a network of creeks. Ranked as England's fastest-growing seaside resort between 1911 and 1951, it was devastated by the North Sea flood of 31 January-1 February 1953, which claimed the lives of fifty-eight islanders and forced evacuation of 13,000 more. Residents were still putting their lives back together on 29 November 1953, when the sea produced another - albeit nonlethal - surprise.

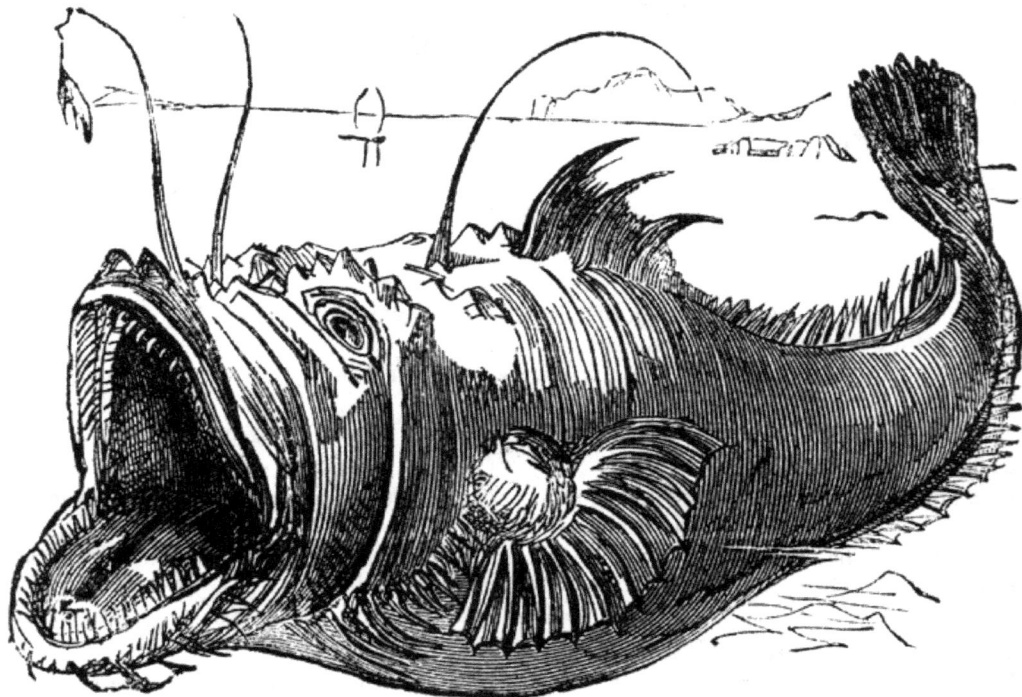

An anglerfish - the Canvey Island monster?

Twelve-year-old Jacqueline Ward found the curious creature, described as a fish with "two feet, each having five toes." It measured thirty inches long and fifteen inches wide, tipping the scales at thirty pounds. It had a "pulpy" head with two protruding eyes, thick brown skin, and a "pear-shaped" tail. Two zoologists from the British Museum of Natural History photographed the thing and, according to Fortean author Frank Edwards, "finally admitted that it looked like nothing they had ever seen before." Half a century later, *Fortean Times* correspondent Gary Hammond suggested that the carcass was "almost certainly" an anglerfish (order *Lophiiformes*), which include three families and various species of pelagic and bottom-dwelling predators. Without the original photos, which have yet to surface, that identification remains tentative.[171]

Girvan, Scotland: 1953

On 21 August 1953 a bristly carcass washed ashore near this fishing village (now a seaside resort) in Carrick, South Ayrshire. According to the *Carrick Herald,* "it was over 30 feet long; had a long neck of about 4 feet, surmounted by what seemed to be a smallish head as though of an animal like a horse or a cow; was covered in coarse grey hair; and had four short stumpy legs!" Its tail alone measured twelve feet in length. Unidentified "experts in London" labeled the beast a "large whale," while other pundits speculated that the Loch Ness Monster had escaped its lake to die at sea. Locals burned the reeking body without preserving any samples of its flesh, but A.R. Waterson of the Scottish Museum and zoologist J.B. Cowey from Glasgow University agreed that the beast was yet another basking shark.[172]

Bondi, New South Wales: 1954

In his 1965 list of stranded carcasses, Dr. Heuvelmans included another oarfish, found by Keith McRae on the beach at Bondi, an eastern suburb of Sydney, sometime in 1954. No further details were provided, and no source offered for the report.[173]

Canvey Island, England: 1954

A second monstrous fish came ashore on Canvey island nine months after the first, on 11 August 1954. Reverend Joseph Overs and his children found it in a tide pool, summoning police to examine the carcass. It measured four feet long, weighed twenty-five pounds, and was in "good condition for examination. Once again, the fish had two short "legs," with five small "toes" arranged in a U-shape. Other features noted in the police report include two large eyes, nostrils, a wide mouth filled with sharp teeth, gills, and "pink skin like that of a healthy pig." Gary Hammond, writing to *Fortean Times,* maintained that the second specimen was yet another anglerfish, species unknown.[174]

Melbourne/Hobart, Australia: 1955/58

Ranked among the most frustrating globsters on record, this puzzling specimen is listed on various Internet websites as possessing "insufficient data" for analysis. Indeed, we may ask whether the "Melbourne-Hobart carcass" ever existed at all. Aside from the obvious fact that Melbourne (the capital of Victoria) is separated from Hobart (Tasmania's capital) by 374 miles, various listings of the incident cannot agree whether it occurred in 1955 or 1958.[175] Efforts to trace it and obtain more information for the work in hand proved fruitless.

Yakutat, Alaska: 1956

Located on the northern coast of the Gulf of Alaska, the Borough of Yakutat sprawls over 9,460 square miles. During the spring of 1956 it produced a strange discovery that was not publicized outside the district until 23 July. On that date, the United Press Association's wire service carried a report headlined "Odd Creature Conceivably Pre-Historic." The story read:

A frighteningly huge carcass, conservatively estimated at more than 100 feet long and 15 feet wide at the broadest visible point, has washed up on the sandy, wind-swept shore of the Gulf of Alaska, 60 miles southeast of here.

The hairy-coated monster has mystified the few persons who have seen it.

Speculation as to what it might have been ranged from an extinct prehistoric beast long encased in a nearby glacier to some warm-blooded sea animal.

At Seattle, Trevor Kincaid, retired University of Washington zoologist, said the description did not fit any prehistoric creature he knew about and that the hair on it precluded its belonging to the living whale or elephant families. He suggested efforts be made to preserve some of its skeletal structure, or the skull and jaws and teeth, or its hide and hair, in efforts to identify it.

A veteran Alaska guide, Earl Lemming, discovered the monster two months ago. Its head measures seven feet in width. The eye sockets, with fragments of decaying flesh still clinging to them, are between seven and nine inches in diameter.

The sockets are approximately 42 inches apart. Reddish-brown hair about two inches long covers its thick, decaying hide. Thick, oily-like blood flowed freely from parts of the flesh when poked with a stick or shovel.

A "flipper" appendage, resembling an elephant's ear, has webbed digits and is about four feet wide and three feet long. The oval upper jaw, with a tusk-like bone, protrudes about 5 feet from the end of the fixed lower jaw.[176]

Life magazine featured the carcass in its issue of 6 August 1956, and the story was still producing echoes four years later. An article in California's *Pasadena Independent*, dated 15 June 1960, reported that a geologist from Denver, Colorado, had viewed the carcass *in situ*, declaring that "It had a head like that of a baby elephant with a snout. It looked like nothing in the world. Nothing I've ever heard of anywhere."[177]

We may certainly forgive a geologist's unfamiliarity with marine wildlife, but what of Dr. Heuvelmans? In his compendium of sea-serpents, he identified the Yakutat beast as a specimen of Baird's beaked whale. The territory fits, but the largest known representative of *Berardius bairdii measured only forty-two feet long, less than half the length of Yakutat's monster.*[178] *Clearly, the case remains unproved.*

Bahia Craker, Argentina: 1957

On 17 December 1957, the Buenos Aires newspaper *La Razon reported that Argentinean naval* personnel had found a "monster" dying at Bahia Craker (Craker Bay) on the coast of

Patagonia's Río Negro Province. It was described as possessing a head that resembled an armadillo's, small eyes without eyelids, and a neck covered by hair three-quarters of an inch in length. A follow-up report, published in the same paper on 30 January 1958, speculated that the still-unidentified beast was a prehistoric mammal, formerly frozen in the Antarctic and lately defrosted before it floated north to strand itself on shore. No further information on the strange case is available today.[179]

New South Wales, Australia: 1959

Sometime in autumn of this year, a "huge serpentine animal" was snared in a fish trap off the northern coast of New South Wales. Fishermen dragged it ashore and stored it in a hangar, where its flesh either decayed and fell away or was removed. Witness A.H.R. Mytleford later viewed the skeleton and notified the *Australian Post,* saying, "Nobody seemed to know much about the object and in fact nobody was interested. But from sketches and illustrations I have seen of sea-serpents, I believe this might be one." Myteford offered £10 for the 18.5-foot skeleton, but the owner refused to sell it. From a photograph, Bernard Heuvelmans identified the skull and spine as a shark's, but could not suggest a species and deemed further research pointless, since "its size was nothing out of the ordinary."[180]

Treasure Island, Florida: 1960

Romantically named Treasure Island is not an island at all, but rather a city in Pinellas County, on Florida's Gulf Coast. Its name derives from one of Florida's perennial land scams - this one dating from 1915, when local property owners tried to boost sales by planting and "discovering" fake treasure chests on the city's beach.[181]

That same beach offered another surprise on 24 March 1960, when William Kanitz found the carcass of a serpentine creature stranded ashore. It proved to be an oarfish, rather than a sea-serpent, however, and interest quickly waned.[182]

Sandy Cape, Tasmania: 1960

The object that gave globsters their name first appeared in August 1960, but did not receive global attention until March 1962. Preceding the discovery, in mid-July 1960, wetern Tasmania was savaged by the worst tropical storm on record. The following month, local rancher Ben Fenton and two of his drovers were rounding up cattle on Sandy Cape, near the Interview River, when they found a large "something" covered with hair like sheep's wool, stranded on the beach. Intrigued by the object, Fenton kept track of it over the next eighteen months, revisiting the site sporadically while tidal action moved the carcass north, some fourteen miles past Sandy Cape. Finally, in February 1962, it came to the attention of Hobart businessman and amateur naturalist G.C. Cramp. He, in turn, alerted state officials and offered to finance a scientific investigation.[183]

What happened next remains a subject of some controversy. A four-man party was dispatched from Hobart, organized by the Commonwealth Scientific and Industrial Research Organisation (CISRO), Australia's national science agency. Team leader Bruce Mollison was described in conflicting media reports as a CISRO staff member, "a zoological

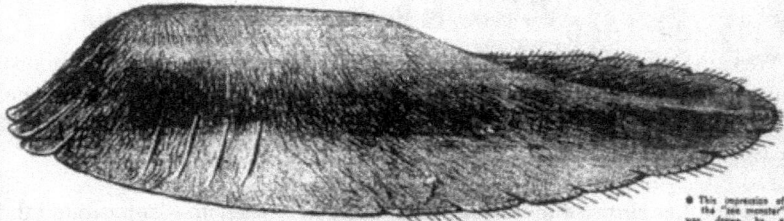

Press coverage of the "original" globster, stranded on Tasmania in 1960.

student," and a person who was "not a science graduate but experienced in some forms of zoology." Assisting him were CISRO member Max Bennett; J.A. Lewis, vice president of the Tasmania Field Naturalists Club; and L.E. Wall. Together, this quartet - described by Malcolm Smith as "two zoologists and two naturalists" - examined the object on 7 March. Mollison reported that the expedition's dogs and horses shied away from the carcass, which smelled like battery acid. Its flesh was hairy, ivory-colored, and too tough to cut with a knife. The remains had no head, eyes, or bones. Beyond that, Mollison observed "five gill-like, hairless slits on each side of the front. There were also four large hanging lobes in front, with a smooth gulletlike orifice between the centre pair. The rim of the hind part of the creature had many cushiony flanges, each carrying a single row of sharp, pencil like spines."[184]

Mollison and company returned on 11 March, armed with an axe and accompanied by a cameraman. Malcolm Smith reports that the photographer shot "hundreds of feet of film" - now lost and presumed nonexistent - while team members hacked off chunks of the carcass for laboratory analysis. Two visits and collection of the samples notwithstanding, on 12

March the *Hobart Mercury* printed an unnamed "scientist's" complaint that "so far no zoologists have looked at it." Questions were raised in Australia's Parliament, and a new team helicoptered off to see the carcass on 16 March. This time, the crew included William Bryden, director of the Tasmanian Museum; John Calaby, senior mammalogist with the CSIRO's wildlife division in Canberra; Eric Guiler, senior lecturer in zoology at the University of Tasmania; and marine biologist A.M. Olsen, senior research officer of the CSIRO's fisheries division in Hobart. With two technicians in attendance, the team comprised "the top names in Australian zoology."[185]

On 18 March the team submitted its report to Sir John Gorton, Australia's Minister for the Navy (and future Prime Minister). It was also published in the *Hobart Mercury* on 19 March, and read as follows:

> The exposed portion of the material was six feet long and two feet wide.
>
> It projected a few inches above the sand surface.
>
> Test holes were dug into the sand around the periphery for several feet to determine the dimensions of the object.
>
> As no solid matter was found in the test holes, we dug around the solid material, passed a rope beneath, and turned it over, thus removing it from the excavation.
>
> When laid out flat, the material was eight feet long, three feet wide, and 10 inches thick at the thickest portion, and from ½ inch to four inches through.
>
> There were a number of irregularly shaped flaps, the juxtaposition of which may have given the impression of clefts, and perhaps the flaps themselves gave an appearance of lobes.
>
> The appearance of the material on its exposed surface was different from that of the buried portions.
>
> In fact, the material is homogeneous in that it consists throughout of tough, fibrous material loaded with fatty or oily substances.
>
> The material has a strong, rancid smell, resembling the higher fatty acids.
>
> The weight of the object was estimated at a few hundred pounds.
>
> The mass was cut through transversely in several places, and particular attention was paid to the flaps.

The material did not contain any bones, spines, or other hard structures.

The hair-like material of the exposed surfaces was merely a consequence of desiccation and leaching of fat-filled fibrous material. Within the body of the material were casual canals, circular in cross section and ½ inch to ¾ inch in diameter.

After examining the solid material, further investigation was made around the site in an attempt to determine the original dimensions of the object.

A few inches below the present sand surface was a layer of sand of variable thickness which had been stained black by organic matter, and had the same strong, rancid smell as the solid matter.

This matter extended eight feet beyond the limit of the solid material in a northerly direction, but to the south and landward sides was only present under the solid material and did not extend beyond its boundaries.

On the seaward side this organic layer extended about 18 feet but we did not consider this distance significant since it follows the natural slope of the beach.

Further investigation was made below the black sand but no solid material was found.

The difference between the size as originally reported and the present dimensions doubtless is due to decomposition and shrinkage.

In view of the fact that this material has been stranded for a long time and is much decomposed, it is not possible to specifically identify it from this preliminary investigation.

Samples have been taken for laboratory comparison by appropriate authorities.[186]

Based on those inclusive findings, Minister Gorton told reporters, "In layman's language, and allowing for scientific caution, this means that your monster is a large lump of decomposing blubber torn off a whale." In fact, the report said no such thing, and CISRO's Bruce Mollison flatly contradicted Gorton, saying that the globster "wasn't fish, fowl, or fruit. It wasn't a whale, seal, sea elephant or squid."

Professor A.M. Clark, at the University of Tasmania, agreed that the carcass "was clearly not a whale," suggesting instead some unknown species of giant ray. Finally, laboratory testing revealed that samples cut from the globster were not blubber at all, but consisted mostly of collagen, the chief component of connective tissue. The last official word simply branded the carcass "a decomposing portion of a large marine animal."[187]

Boulmer, England: 1961

Boulmer - pronounced "Boomer" - is a coastal village in Northumberland, located on the North Sea, east of Alnwick. It boasts a Royal Air Force station that is is currently home to Aerospace Surveillance and Control System Force Command, the School of Aerospace Battle Management, and No. 202 Squadron RAF search and rescue. On 21 March 1961 a globster beached at Boulmer, but after brief excitement it was found to be another basking shark.[188]

Barra, Outer Hebrides: 1961

A carcass omitted by Dr. Heuvelmans from his 1965 list of globsters came ashore on Barra, southernmost island of Scotland's Outer Hebrides, in July 1961. On 18 July, the *Scottish Daily Express* published a photo of the specimen, appearing to display a long neck, but the accompanying article revealed that Peter Usherwood, a zoologist from Glasgow University, had identified the "monster" as a male beaked whale (family *Ziphiidae*). Sadly, decomposition had advanced to far for Usherwood to say which of the family's twenty-one species it represented.[189]

Saint-Jean-des-Monts, France: 1961

Another rotted basking shark carcass came ashore on 24 December 1961, at Saint-Jean-des-Monts in the Vendée department of Pays de la Loire, a region in western France. It caused the normal uproar, soon subsiding when the carcass was identified.[190]

Ucluelet, British Columbia: 1962

On 14 April 1962 the Vancouver *Sun* reported the discovery of a fourteen-foot carcass with a head like an elephant's, floating offshore from the village of Ucluelet, on the northern shore of Barkley Sound. A fisherman named Simon Peter had hauled it ashore for closer examination, leading to tentative identification as a decompsed elephant seal. Authors Paul LeBlond and Edward Bousfield accept that verdict, without citing further details.[191]

Coney Island, New York: 1962

This case appears to be an Internet hoax, but I include it here for the sake of full coverage. According to a website called "The Incredibly Strange Wildlife Garden," the rotting carcass of a giant octopus was beached near Coney Island sometime during 1962. According to author Stanton Wood, the hulk measured seventeen feet long and weighed 450 pounds. It was shipped off for study at Cape Cod's Woods Hole Oceanographic Institute, where it was identified as "a specific species of giant octopus" possessing only seven arms, then "the incident [was] forgotten." More surprisingly, it was also ignored by the *New York Times,* which logged no strandings for the year but did report the debut of a "giant" octopus called "Handy Andy" at Coney Island's aquarium, on 28 March 1962.[192]

As if the tale so far was not bizarre enough, Wood buttressed it with a claim that a giant cephalopod attacked the fishing boat *Bonny Harlot* on Long Island Sound, sometime during 1987. In yet another case that managed to elude the *Times,* this Kraken-like monster supposedly wrapped three eighteen-foot arms around the vessel, retreating only after crewman Lyle Cavendish severed one arm with a fire axe. The relic was sent to Woods Hole (of course!), where author Wood tells us it helped identify "a heretofore undiscovered new species of octopus." Seaman Cavendish promptly retired, to become a preacher in rural Idaho, while - Wood claims - the original 1962 stranding inspired a (nonexistent) film titled *Octopus in My Soup*.[193] In fact, the hoax leaves a sour taste in our mouths, and we shall now move on.

Île du Levant, France: 1963

Our next globster surfaced on 9 July 1963, beached on this island off the Mediterranean coast of France, known as a resort for nude bathers. The carcass was quickly identified as that of an oarfish, but at least it existed, unlike the manufactured urban myth from Coney Island.[194]

Malibu, California: 1963

While oarfish are considered relatively rare, another managed to strand itself near Malibu, California, on 24 September 1963. Carole Richards found the carcass, while walking her dog on the beach. It measured eighteen feet in length and is now displayed at the Los Angeles County Museum of Natural History.[195]

Whidbey Island, Washington: 1963

Whidbey Island, at the north of Puget Sound, is one of nine islands comprising Island County. The only well-known incident in Whidbey's history occurred on 8 December 1984, when FBI agents killed Robert Jay Mathews - leader of a neo-Nazi terrorist cult, "The Order" - at his fortified bunker near the town of Freeland. For purposes of our study, however, a more significant event took place on 29 September 1963.

That afternoon, a mere five days after the Malibu oarfish appeared, Ruth Cobert found a rotting carcass half-buried in sand on Whidbey Island's Sunset Beach. It measured twenty-five feet long, with a twenty-inch skull resembling that of a horse. The creature's spine was six inches thick at the skull, dwindling to two inches in diameter at the tail. The corpse also revealed some unidentified "cartilaginous material." Dr. A.D. Welander, from the Fisheries Department at the University of Washington, examined photos of the carcass and identified it as a basking shark.[196]

Skaket Beach, Massachusetts: 1964

Skaket Beach is located on Cape Cod, near the town of Orleans. On 24 December 1964 a local newspaper, the *Cape Codder,* reported that fisherman Elmer Costa had found an eighteen-foot carcass stranded on the beach. According to that article, "It lay with its long serpentine tail pointed slantwise up the beach, its tiny head (perhaps the skeletal size of a raccoon's) at the end of four feet of snaky joints and bones, looking for all the world like a baby pterodactyl, widening in the middle into a structure that could have been wings gone swollen by decay." In fact, however, Dr. Richard Backus from the nearby Woods Hole Oceanographic Institute quickly arrived and put that fantasy to rest, naming the carcass as another basking shark.[197]

Muriwai Beach, New Zealand: 1965

Muriwai and its eponymous beach are located on the west coast of New Zealand's North Island, facing the Tasman Sea, ten miles west od Kumeu and twenty-six miles northwest of

Auckland. In March 1965, beach strollers found a globster on the shore that measured thirty feet in length and eight feet tall, covered in "hair" four to six inches long. Dr. J.E. Morton, a zoologist dispatched from Auckland University, examined the remains and told reporters, "I can't think of anything it resembles." There the matter rests, without a photograph or sample of its flesh to analyze.[198]

Lac Sinclair, Québec: 1966

During the summer of 1966, teenager André Arsenault and an unnamed friend set out to explore the swamps surrounding Lac Sinclair, west of Wakefield in Canada's Ottawa Valley. In a nearby river, they found an object "floating and wallowing in two or three inches of water." Arsenault later described it for authors Michel Meurger and Claude Gagnon.

It was obviously a section of a cylinder; light browny yellow, 24 inches long and 10 inches in diameter. We raised it with an oar.It was heavy. It didn't have the texture of a fish. And there were no fins. It was like a tube. You could say it was like a clean-cut sausage, but what could have cut this piece? My friend, who was an habitual fisherman, had never seen anything like this. He said that it resembled part of an enormous *serpent.* Even the colour wasn't that of fish flesh. It couldn't have been there more than a week, because the flesh was still firm. If it was part of a serpent, like I think, *it must have been an enormous one.* In the same lake there are many small snakes and tortoises.[199]

Whatever the mini-globster may have been, it remains unidentified.

Bandon, Oregon: 1967

An equally enigmatic case hails (allegedly) from Bandon, located at the mouth of the Coquille River in Coos County, Oregon. I first encountered mention of this case online, cited in an article by journalist J.D. Adams. That piece provided no date or details, beyond the statement that "a sea monster" more than twelve feet long, with "a hairy, cow-like body," had washed ashore near the small town of two thousand residents. Upon contacting Adams by email, I learned that his source was Marge Davenport's book *Afloat and Awash in the Old Northwest,* published in 1988 by now-defunct Paddlewheel Press. Davenport disposed of the incident in one sentence, again omitting any reference to dates or other details, but cited as her source an article from Portland's *Oregonian* newspaper, written by Peter Cairns on 24 September 1967. That article, in turn, simply refers to "Bandon's mini-monster, a 12½-foot animal with a bulbous nose and a cow-like body covered with brownish hair."[200]

Tecoluta, Mexico: 1969

Tecoluta is a village on the Gulf of Mexico, in the state of Veracruz. In March 1969, a thirty-five-ton carcass came ashore there, sparking global publicity. It was described as having a "serpent-like body covered with hard jointed armor," while a ten-foot tusk of bone weighing about two thousand pounds protruded from his head. Based on the sole surviving photograph, that head did not resemble any species known to live in modern seas. A wire story from United Press International claimed that several unnamed biologists initially thought the creature was a narwhal, then reconsidered, deciding that "they could not match it with any sea

The enigmatic Tecoluca carcass.

creature known to man."[201]

Certainly, they had been wrong in their narwhal diagnosis, since those cetaceans rarely exceed, 3,500 pounds, with the longest tusks on record weighing less than thirty pounds. A second panel viewed the carcass on 20 April, declaring it to be a rorqual whale (family *Balaenopteridae*). They did not choose among the familiy's nine species, but none bear armor or tusks. The scientists proclaimed themselves satisfied and recommended immediate disposal of the corpse, but Tecoluta's mayor insisted on keeping the malodorous relic ashore as a tourist attraction until it decayed beyond recognition and was finally reclaimed by the sea. Its true identity remains a point of controversy.[202]

Mann Hill Beach, Massachusetts: 1970

In November 1970 a twenty-foot carcass washed ashore at Mann Hill Beach, in Plymouth County. Observers estimated that it weighed fourteen to nineteen tons. It seemed to possess a serpentine neck and flippers, prompting one witness to dub it "a camel without legs." Others compared it to a plesiosaur, but it proved - once again - to be merely a decomposed basking shark.[203]

The Mann Hill Beach globster.

Sandy Cape, Tasmania: 1970

Few persons can claim discovery of a single globster, but rancher Ben Fenton found his second in November 1970, near the site of his first find, ten years earlier. This one lay half-buried in sand on a beach thirty miles south of Temma. The blob was somewhat larger than its predecessor, at nine feet nine inches long and nearly four feet wide, covered with the same "woolly" growth as the 1960 specimen. Having delivered their vague and inconclusive diagnosis eight years earlier, no scientists saw fit to visit Sandy Cape a second time.[204]

Loch Ness, Scotland: 1972

On 31 March 1972, staff members from the Yorkshire Zoo informed reporters that they had found the remains of a large unknown animal beached on the eastern shore of Loch Ness, near Foyers. They removed the carcass - packed in ice, aboard a truck - for scientific study back in England, while some locals volubly protested that if Nessie had been found, it should remain in Scotland. Aroused by the controversy, police stopped the truck and recovered the carcass - which proved to be an elephant seal with its whiskers removed and cheeks padded to alter its appearance. The "discovery," as finally revealed, was an joke planned to break on 1 April - April Fool's Day.[205]

Durgan Beach, England: 1976

The name "Durgan Beach" serves two separate shingle-and-pebble beaches accessible by footpaths from the hamlet of Durgan, on the Helford River in south Cornwall's parish of Mawnan. The Helford, in turn, is not a true river, but rather a ria: a flooded valley fed by seven creeks that ultimately flow into the English Channel. Late in January 1976, the Channel gave up a carcass whose identity is claimed by some authors to be a mystery, although they should know better.

An early, anonymous report of the case described the "strange (and so far unidentified) carcase...discovered on Durgan Beach, Helford River, by Mrs. Payne of Falmouth." A quarter-century later, author Paul Harrison still claimed that the globster "baffled those who saw it, since it was of no discernible shape. Unfortunately, it was washed back out to sea before it sould be officially examined."[206]

Well, not quite. A reading of the *Falmouth Packet* (5 March 1976) reveals, first, that the Durgan Beach carcass consisted of skeletal remains, and that a teenaged naturalist had tentatively matched them to a partial skeleton he found on Cornwall's Prisk Beach, close to Durgan, shortly after Christmas 1975. The Prisk Beach relic included a skull, which was preserved at a local school, later examined by Jonathan Downes of the CFZ and Dr. Karl Shuker. Both agreed that the creature had been a cetacean, a verdict confirmed by experts at the Plymouth Aquarium. They opined that the specimen was "probably a pilot whale," and undoubtedly one of the smaller toothed whales (suborder *Odontoceti*), including dolphins, beaked whales, sperm whales, and others.[207]

South Pacific Ocean: 1977

One of history's most controversial globsters was snared in a net by the Japanese fishing

Three views of the *Zuiyo Maru* carcass.

vessel *Zuiyo Maru* on 25 April 1977, some thirty miles offshore from Christchurch, on New Zealand's South Island. Hauled up from a depth of 975 feet, the rotting carcass measured thirty-three feet long and weighed about four thousand pounds. Crewmen snapped photographs of the apparent long-necked creature, flippers dangling, and Michihiko Yano, the ship's assistant production manager, plucked forty-two "horny fibers" from the left-front flipper. Yano also sketched the creature, though published reports disagree as to whether he made the drawings while the carcass lay on deck, or after the *Zuiyo Maru* returned to Japan. Finally, on the captain's orders, the globster was dumped overboard to prevent contamination of the ship's edible cargo.[208]

Photos of the *Zuiyo Maru* carcass were broadcast worldwide, prompting comparisons of the corpse to a plesiosaur, a giant turtle or sea lion, and a rotting whale. Aboard the ship, meanwhile, Yano washed his fibrous samples in an antiseptic solution of sodium hypochlorite, later delivering them to biochemist Shigeru Kilmora at Tokyo University. In due time, Dr. Kilmora reported that the fibers contained a protein called elastodin, found only in sharks. Other analysts, working only from the shipboard photographs and Yano's sketch, declared that

(a)   "the anterior limb or fin appears to be articulated at a right angle to the shoulder. Such state of articulation is indicative of a shark";

(b)   "we can clearly distinguish the base of a dorsal fin, though it had slipped from the mid-dorsal line";

(c)   "There were mycomata in the dorsal muscles" (strong connective tissue between muscle segments, found in sharks but not in reptiles); and (d) that "lack of [a] neural spine is consistent with selachians." With regard to plesiosaurs, all agreed that the globster's skull and spine were inconsistent with known species, while the carcass had "pectoral and dorsal fins with fin-rays characteristic of fish."[209]

Thus, the riddle was definitively solved. Or, was it?

Opposing the shark diagnosis for religious reasons, members of the Christian-Creationist community have highlighted various perceived discrepancies from the official Japanese reports. Among them, we read that -

(1)   "The surface of the body was whitish and covered with dermal fibers which intersecting each other like in whales and other mammals but were not weak as in fish. There were thick, white, fat-like tissues on the back, and reddish muscles were seen running longitudinally beneath the white tissues.... The head was said to have been hard, exposing the cranium, and not shark-like.... Unlike sharks, in which the nares are situated in the lower surface of the skull, the carcass had nares at the front end of what remained of the cranium.

(2)   "The putrefactive smell was not like that of teleostean fishes or sharks, but resembles that of marine mammals. The carcass did not smell of ammonia, which is a characteristic feature of shark flesh. An explanation for this could be due to the extent of skin loss and decomposition, and therefore allowing the ammonia from the carcass to be washed out by the sea."[211] Creationsts ask why that should be the case, when "all other" shark carcasses reek of ammonia.[212]

(3)   "The strongest argument opposing the shark theory comes from Yano's observation that the carcass was covered by a fat-like sticky substance. Sharks do not have a thick layer of fat under the skin."[213]

For those and other reasons - red muscles observed along the globster's spine, four large limbs with "the posterior pair...almost equal to the anterior one in size," and rigid bones detected in one flipper when Yano stepped on it in passing - Creationists assert that the carcass was in fact that of a prehistoric marine reptile, thus validating their doctrinal claim of a "young Earth," some 6,000 to 10,000 years old, in which early humans (formed by God from handsful of dust) shared the planet with dinosaurs until Noah's flood made the great land-dwelling reptiles extinct. They cite Professor Tokio Shikama, a paleontologist at Yokohama National University, who declared: "Even if the tissue contains the same protein as the shark's, it is rash to say that the monster is a shark. The finding is not enough to refute a speculation that the monster is a plesiosaur." How could that be? Dr. Shigeru Kimura, a biochemist at the Tokyo University school of fisheries, opined, "Among fish, it is known that only sharks and rays have the type of protein called elastodin. But as for reptiles, I do not think there is relevant data, even abroad."[214]

And as for prehistoric reptiles, known only from fossilized remains ....

Creationist author Malcolm Bowden writes that:

> [t]o admit that there were plesiosaur-type animals still living today would cause considerable consternation to evolutionists. As you go back in time there are an increasing number of reports of dinosaurs

both on land and in the sea. Indeed, in Medieval times they were almost commonplace. This indicates that they were very numerous not all that long ago. This does not fit with the evolutionary timescale that they all disappeared about 65 million years ago, and after such a long time, none should exist. It is for this reason that reports of any sightings are ignored in the scientific world and evidence such as this carcass is rapidly contradicted. [215]

And while, strictly speaking, there were no aquatic dinosaurs - the term properly applies only to land-dwelling reptiles living from the late Triassic period through the end of the Cretaceous - he is undoubtedly correct about the impact science would endure if prehistoric megafauna surfaced, alive and well, in the twenty-first century. Undercutting that argument is the fact that most cryptozoologists suspect any unknown "sea-serpents" are likely large mammals, not reptiles at all.

Be that as it may, arguments surrounding the *Zuiyo Maru* carcass continue, at least along sectarian lines.

Kotu Beach, Gambia: 1983

The first reported globster of the 1980s also ranks as one of the stranger on record. Before addressing the globster itself, however, we must first resolve a minor problem of geography. Published reports of the incident universally refer to "Bungalow Beach, Gambia," but in fact, no such beach appears to exist. There is a highly-rated Bungalow Beach *Hotel,* located on Kotu Beach near Gambia's largest city, Serekunda. A comparison of Kotu Beach's location with a map of "Bungalow Beach" published by authors Loren Coleman and Patrick Huyghe indicates the two to be identical.[216]

On 12 June 1983, while vacationing at Kotu Beach with his family, teenage "wildlife enthusiast" Owen Burnham found a strange creature stranded dead on the sand. Caught without a camera, he memorized details of the carcass, later saying that it measured fifteen to sixteen feet in length, with black hide on its dorsal surface and lighter skin below. Its long jaws contained eighty teeth, with two nostrils on the snout. The body had two pairs of flippers, with one of the hind pair nearly severed, but otherwise displayed no significant damage. That quickly changed when local natives dismembered the carcass and carried its parts off to fill cooking pots.[217]

Corresponding with Dr. Karl Shuker in 1986, Burnham added further observations: "There was no sign of any blowhole....The creature can't have been dead long, because its eyes were clearly visible and brown although I don't know if this was due to death. (They were not protruding.) The forehead was domed though not excessively. (No ears.)" [218]

Shuker noted that, since all known cetaceans have dorsal blowholes and a single pair of flippers, located forward, the creature - nicknamed "Gambo" - could not have been a dolphin or small whale. He concluded that the only species matching Burnham's description were both prehistoric and long presumed extinct: a short-necked relative of the plesiosaurs, called a pliosaur (family *Pliosauridae*), or an early marine crocodile (suborder *Thalattosuchia*). Coleman and Huyghe disagree, classifying Gambo as a "mystery cetacean."[219] Thus far, the mystery remains unsolved.

Sakhalin Island, Russia: 1986

The decade's next carcass came ashore sometime in 1986, from the Strait of Tartary, stranding at Gornozavodsk on Sakhalin Island (not to be confused with the *other* Gornozavodsk, located 119 miles northeast of Perm, on the Russian mainland). The body measured twenty-six feet long, with long flattened jaws more than three feet in length, containing visible teeth. The tip of its tail had been severed, leaving vertebrae exposed. Sergei Litvinyuk, attached to a geological party working in the area, viewed and photographed the carcass in September, initially identifying it as beaked whale, species unknown. He collected no samples, but noted "one or two pairs" of teeth at the tip of the snout, "resembling incisors of hoofed animals."[220]

Somewhat curiously, Litvinyuk later changed his mind, collaborating with Grigory Panchenko to write a report that re-identified the creature as an archeocete ("ancient whale," suborder *Archaeoceti*), from the prehistoric group including the basilosaurids, ancestors of modern whales. Presumed extinct since the Oligocene epoch, archaeocetes are not expected in Russian waters, or anywhere else on modern Earth.[221]

Russian cryptozoologist Dmitri Bayanov translated the Litvinyuk-Panchenko report into English in 1989, and a copy found its way to author Darren Naish via Richard Freeman, of the CFZ. Together, they concluded that the Sakhalin carcass was "almost certainly" a specimen of Baird's beaked whale, rather than a survivor from the distant past.[222]

Mangrove Bay, Bermuda: 1988

In May 1988, fisherman Teddy Tucker discovered a globster at Mangrove Bay, on the west coast of Somerset Island, Bermuda. Somerset itself lies to the far west of this British overseas territory, comprising roughly half of Sandys Parish. Tucker later told author Richard Ellis that the object was eight feet long and "2½ to 3 feet thick...very white and fibrous...with five 'arms or legs,' rather like a disfigured star." He pegged its weight at close to a ton, noting that three persons working together could not turn it over. While it had no bones or cartilage, the blob was "very dense and solid." When he tried to slice it with a knife, Tucker said, "it was like trying to cut a car tire." Nonetheless, he succeeded in obtaining samples before the Atlantic Ocean reclaimed its strange prize.[223]

Photos of the "Bermuda Blob" were sent to various experts, including Forrest Wood, Roy Mackal, and Clyde Roper, the Smithsonian's top expert on *Architeuthis*. In April 1995, author Sidney Pierce and three colleagues examined samples of the Bermuda carcass, concluding that it was collagen "from the skin of a poikilothermic vertebrate...either a large teleost [ray-finned fish] or an elasmobranch." French cryptozoologist Michel Raynal countered with the

OPPOSITE: Teddy Tucker examines the 1988 "Bermuda Blob."

observation that "[n]o known fish or reptile has a skin thick enough to possess such a mass of collagen." Nine years later, however, Pierce and five coauthors revisited the sample, declaring that the Bermuda Blob and five other globsters were "nothing but whales." The obvious contradiction remains unresolved - and was in fact compounded when Pierce and company reversed themselves in 2004, identifying the Bermuda globster as whale blubber.[224]

Godthåb, Greenland: 1989

Various Internet websites offer lists of carcass strandings throughout history, including several that mention a "Godthaab Globster (1989)" without any further information or source citations. "Godthaab" is, in fact, *Godthåb* ("good hope"), the capital of Greenland, better known to that nation's inhabitants as *Nuuk* ("the penninsula"). One site adds that the globster was stranded in summer, but that failed to advance my search for details.[225] At press time for *Globsters,* no more data was available.

Benbecula, Outer Hebrides: 1990

Benbecula, in Scotland's Outer Hebrides, is known in Gaelic poetry as *An t-Eilean Dorcha* - "the dark island." In 1990, sixteen-year-old Louise Whitts found something on the island's beach and posed beside it for photographs. When she returned to study it again, next day, the carcass had vanished, presumably swept back to sea by the tide.[226] As Whitts described the object:

> It had what appeared to be a head at one end, a curved back and seemed to be covered with eaten away flesh or even a furry skin and was about 12ft. long. It smelled absolutely disgusting, but the weird thing was that it had all these shapes like fins along its back... like a dinosaur or something.[227]

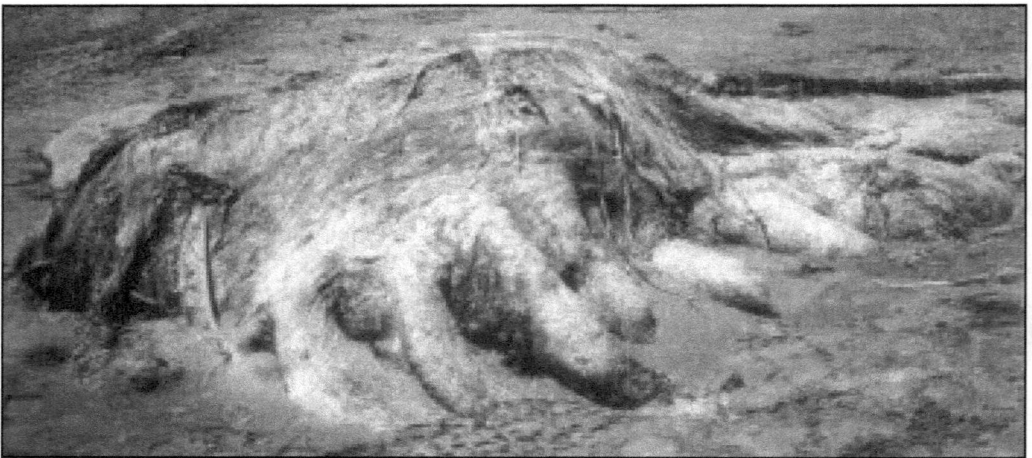

A photo of the Benbecula globster, donated by Louise Whitt to the Hancock Museum in Newcastle upon Tyne.

Bearing in mind that dinosaurs had no fins, the observation solved nothing. Likewise, when Whitts rediscovered her photos of the globster in August 1996, donating them to the Hancock Museum in Newcastle upon Tyne, no answers were forthcoming. As noted in *Fortean Times,* "none of the botanists, zoologists and marine biologists who had seen the pictures could throw any light on the puzzle."[228] So it remains, to this day.

Cape Meares, Oregon: 1990

In June 1990, Michael Cenedella and companion Joanne Rauche went strolling on Cape Meares, at the southern end of Oregon's Tillamook Bay, and found a strange carcass beached before them. Cenedella paced off its length at thirty-three feet and snapped several photos, which later appeared on various Internet websites. An unidentified spokesperson for Oregon State University's Hatfield Marine Science Center suggested that the beast might be a decomposed gray whale (*Eschrichtius robustus*), which may grow to fifty feet in length but does not possess a dorsal fin or an apparent leg resembling those depicted in Cenedella's snapshots. Cenedella himself had seen several gray whales, and insisted that "[n]one looked remotely like this. The grays don't taper nearly so much at the tail and don't taper at all at the head. The heads are massive." Cenedella also noted a bystander's opinion that the carcass's "bent leg" might be, in fact, "a grotesque penis."[229] No samples of the creature's flesh were taken for analysis, and so the mystery remains unsolved.

Bermuda: 1995?

Several Internet websites - led by Wikipedia, with others generally plagiarizing its text verbatim - claim that a second globster washed ashore on Bermuda in 1995. Neither the place nor date is ever specified, but the same websites claim that "Bermuda Blob 2" was analyzed by Sidney Pierce and five colleagues in 2004, identified as adipose tissue from a decomposed whale.[230]

A problem rears its head, however, when we seek more information on the case at hand. While researching *Globsters,* I contacted the Bermuda National Library, in hopes of pinning down the date of the globster's discovery. Instead, librarian Ellen Hollis informed me that loocal newspapers contain no mention of a carcass found at any time in 1995. Neither do library archives at the Bermuda Aquarium, Museum and Zoo. In fact, it seems that "Bermuda Blob 2" did not arrive until January 1997, in an incident described below - and listed by various website's as representing Bermuda's *third* globster. Finally, the Pierce report itself confirms that the second Bermuda globster was beached in January 1997. Wikipedia's confusion may have arisen from the fact that Pierce et al. incorrectly claim that "Bermuda Blob 1" was discovered in 1995, rather than 1988.[231] This oft-repeated incident must now be stricken from the record.

Loch Ness, Scotland: 1995

On 15 June 1995, self-identified "forensics investigator" William McDonald announced the discovery of a large, peculiar tooth or fang at Loch Ness. Two unnamed "American students" supposedly found the tooth in March 1995, embedded in the carcass of a half-eaten deer near the loch. Other unnamed witnesses, allegedly interviewed by McDonald in December 1994,

The Loch Ness "tooth" -a literary publicity stunt.

described an eel "as long as a lorry" crawling alone the lake's shoreline, and McDonald himself claimed discovery of drag marks indicating a creature fifty-five feet long and weighing some eight tons.[232]

Alas, the whole thing proved to be a hoax - or, more precisely, a commercial publicity stunt, timed to accompany the release of a horror novel, *The Loch,* written by best-selling author Steve Alten. The "tooth" in question proved to be an antler from a muntjac deer, while McDonald's other claims went up in smoke. Author Mike Dash later referred to an apparent separate case, involing "rumors that remains of an 18ft eel were found in a water inlet at Foyers aluminum works around 1995" and resurfaced in a June 1997 BBC2 documentary, "but have been denied."[233] Research for *Globsters* revealed no further details on that incident.

Pulau Langkawi, Malaysia: 1996
Pulau Langkawi is the largest of four inhabited islands among ninety-nine that comprise the Langkawi archipelago, located in the Andaman Sea, off the west coast of Peninsular Malaysia. On 27 May 1996, local fishermen netted the skeletal remains of an unknown creature near Pulau Langkawi, hauled up from a depth of 175 feet. Various reports described the skeleton as being ten to twenty-five feet long, while photos published in *New Straits Times* suggest a

length somewhere between those two extremes, compared to humans in the frame. Early reports claimed that the fishermen, fearing they'd pulled in a dragon, quickly returned their catch to the sea. Later accounts said it was purchased by a village "healer," who deodorized its stinking flesh with bleach, then rendered it for use in potions. Curiously, the initial photos show no flesh to speak of on the carcass.[234]

The surviving photos, published worldwide, sparked controversy concerning the creature's identity. Spokesmen for the Malaysian Fisheries Department confessed that they had "never seen anything like it." Dr. Karl Shuker examined the photos and said, "The teeth and vertebrae seem mammalian, yet the head seems reptilian." Fishery taxonomist Mohamed Azmi Ambek subsequently identified the skeleton as that of an orca, whose global range includes Malaysian waters.[235] At a glance, the "dragon" skeleton does strongly resemble an orca's, but lacks the whale's prominent ribcage.

Block Island, Rhode Island: 1996

Less than three weeks after the Malaysian "dragon" surfaced, in mid-June 1996, fishermen Gary Hall and J.T. Pinney hauled in another carcass while trawling from the *Mad Monk,* near Block Island, thirteen miles south of Rhode Island's Atlantic coastline. Contradictory published reports claim that the catch was made on 17 and 20 June, pegging the length of the serpentine creature at thirteen to fourteen feet. The creature had a foot-long snout, protruding from "a narrow head with vacant eye sockets and weird whiskers."[236]

Hall and Pinney displayed their odd catch near the dock for the Port Judith ferry, where gawkers studied it over the next two days. Lee Scott, a state park biologist, then took the carcass home and placed it in his freezer, planning to deliver it for study at the National Marine Fisheries Service facility in Narragansett, Rhode Island. Before he could follow through, however, unknown thieves absconded with the carcass and it vanished forever, leaving locals and tourists to celebrate the missing "Block Ness Monster" with T-shirts sold for eighteen dollars apiece.[237]

What was the Bock Island creature? Young author Ben Roesch identified it as a basking shark, but Lee Scott disagreed, noting that the twelve-inch snout was twice the length of that normally seen on *Cetorhinus maximus*. Other pundits theorized that it may have been a sturgeon. The question remains unanswered.[238]

Cape Hatteras, North Carolina: 1996

On 6 September 1996, Hurricane Fran ravaged the Eastern Seaboard of the United States, claiming thirty-seven lives and causing $5 billion damage in fourteen states. Nearly half of that property damage was suffered in North Carolina, but the storm also delivered a peculiar object to a beach on Cape Hatteras National Seashore.

Teenager Brad Abrams and five friends from Manteo, in Dare County, drove to Cape Hatteras as soon as the storm dissipated, dodging clean-up chores at their respective homes in favor of surfing and beachcombing. En route to the beach, they joked about finding pirate treasure unearthed by the gale, but instead, as Abrams later said, "What we did find was this huge stinking mess."[239]

Specifically, they found a rotting globster twenty feet long, six feet wide, and over four feet thick at its middle. "I thought maybe it was a dead whale or octopus that had been killed by the waves," Abrams told an Internet scribe, "but it didn't look like anything. I mean it had no head or tail. There was these tentacle like things coming out from it but they just sort of came out at random points. It was like God's kid made some sort of animal while dad wasn't looking and it came out all wrong."[240]

By the time Abrams paid another visit to the site, on 17 September, the globster was gone, presumably washed back to sea. A year after the fact, Abrams showed his photos to a biology teacher at the University of North Carolina in Wilmington, who in turn reportedly sent them to the "International Cryptozoology Association," location unknown. Despite Internet citations of an ICA newsletter, allegedly published in November 1997, the organization - assuming it ever existed - proved terminally elusive during my research for *Globsters*. The International Society of Cryptozoology ceased operations in 1996, so Abrams's report cannot be explained by a simple confusion of names.[241]

Nantucket Island, Massachusetts: 1996
According to Dr. Sidney Pierce and his colleagues, a globster of unspecified dimensions and weight washed ashore on Nantucket Island "sometime in November 1996." A search of local newspaper records, conducted on my behalf by librarians at the Nantucket Atheneum, reflect no such incident, nor did the discovery rate mention in the *New York Times*.[242]

Nonetheless, something did come ashore on Nantucket at some point prior to June 2004, when Pierce et al. published their study of that globster and five others. Samples of the carcass were collected, photographed and frozen by Nantucket's shellfish warden. The flesh yielded amplifiable DNA which was 99 percent identical to that of a finback whale (*Balaenoptera physalus*).[243]

Claveria, Masbate, Philippines: 1996
CFZ correspondent Markus Hemmler reports that our next globster beached on 24 December 1996, "on beach villa Rico near the town of Claveria, located in the Philippine province of Masbate." Claveria is, in fact, a fourth-class municipality in Masbate Province, located on Buris Island. It is subdivided into twenty-two *barangays* (villages), none actually named Claveria. To confuse matters further, there are two more municipalities called Claveria, found in the provinces of Cagayan and Misamis Oriental.[244] My efforts to locate "beach villa Rico" proved fruitless.

Be that as it may, there is no question that something washed ashore on Buris Island. Published reports describe the carcass as twenty-six feet long, resembling an "eel-like creature with the head of a turtle." Hemmler reports that the *Philippine Star* published a photo of the creature's skull, vertebrae and limbs, but said photo has apparently been lost, and the newspaper's online archives include no stories prior to 2000. Various unnamed experts examined the photo and samples of flesh from the carcass, without producing a confirmed identification.[245]

Of those who ventured suggestions, one - a Canadian missionary - thought the carcass was a plesiosaur's. Zoologist Perry Ong, from the University of the Philippines, described the creature as an "eel-like fish...It must be an ancestral or primitive fish. It had fins. But if it is a fish, where are the ribs? It is not a mammal." Karl Shuker disagreed, suggesting that the carcass might belong to an orca. Ben Roesch and Markus Hemmler cast their votes for a basking shark. Hemmler cites as his authority a report published by the Commonwealth Scientific and Industrial Research Organisation in 2005, referring to skeletal remains of a basking shark stranded at "Iceland Burias, Masbate" in December 1996.[246] While there is no "Iceland" in the Philippines, it would be strange indeed if this were not the selfsame globster.

Bermuda: 1997

Described on various Internet websites as "Bermuda Blob 3," this globster was actually the *second* carcass stranded on Bermuda. It came ashore at South Beach, Warwick Parish, on 11 January 1997, and was first seen by a teenage beachcomber. Dr. Wolfgang Sterrer, curator of the Bermuda Aquarium and Zoo, turned up to see the carcass on 12 January, with scientists from the Bermuda Biological Station. They estimated that the carcass weighed a thousand pounds, Sterrer describing it as "definitely a piece of a large marine creature." His "gut instinct" suggested a whale, but Sterrer advised caution, since "the gut instinct is not always right." Despite complaints from nearby residents, Sterrer and company left the reeking relic where it lay for several days, while they collected samples of its flesh for DNA analysis and other tests. On 14 June, aquarium spokesman James Conyers opined that flesh from the carcass "smelled, looked and was the texture of decomposed whale blubber." Sidney Pierce and his five coauthors confirmed that judgment in 2004.[247]

Al Fintas, Kuwait: 1997

Al Fintas is located on the Persian Gulf, south of Kuwait City. On 8 August 1997 a municipal employee, Mohammad Yousef Obaid, found a strange carcass floating "face down," a few yards offshore, and hauled it up to the beach. "First," Obaid told the *Jordan Times,* "I thought it was a dead human body and wanted to call police. But then I thought let me determine for myself what it was and so I turned it face up with a stick. I was shocked to see that the creature, or whatever, had a strong resemblance to humans. I could see that the body was disintegrating but could distinctly spot the strange skull, remnants of eyes, ears and mouth, spinal cord and pelvis."[248]

Media reports described the carcass or skeleton as four feet ten inches long, with "a big mouth, two nostrils, two eyes and two ears which jut out of the skull resembling a Chinese dragon or even the 'devil' as portrayed by artists." It broke into three pieces as Mohammad Obaid dragged it ashore. He took the skull with its attached spinal column home, depositing them in his freezer, then found the rest gone by the time he returned to the beach. His calls to various Kuwaiti research centers ended, Obaid said, when "no one seemed to take me seriously."[249]

The fate of the Al Fintas "fishman" remains unclear. The *Jordan Times* speculated that "[i]t could be a sea monster, a rare species of fish, a creature from outer space, or a total hoax altogether." Obaid countered the latter suggestion by saying, "It is no hoax. I think it will be a big shame if such a rare find goes uninvestigated." Thus far, there has been no investigation, although Ben Roesch viewed a photo of the globster and opined that it might be a cownose ray (*Rhinoptera bonasus*).[250]

Wanganui, New Zealand: 1997

On 14 October 1997, CNN reported that a carcass resembling "a lump of rotting carpet" had washed ashore from the South Taranaki Bight near Wanganui, on New Zealand's North Island. According to that report, the globster smelled foul, while "[a] hairy fiber grew over its whitish flesh, and it has large, paddle-like tentacles." Steve O'Shea, from the National Institute of Water and Atmospheric Research, sniffed the carcass and pegged it as a rotting sperm whale. Other unnamed experts were examining the globster, "hoping once and for all to unravel the mystery behind it," but no follow-up report was forthcoming.[251]

Four Mile Beach, Tasmania: 1998

Our next globster turned up at Four Mile Beach, north of Zeehan on Tasmania's west coast, in early January 1998. The *Illawarra Mercury*, published in Wollongong, New South Wales, described it first, on 9 January, and published several photos. The carcass measured sixteen to twenty feet in length and six feet wide, with an estimated weight of three to four tons. It was hairy and malodorous, with at least six "flipper-like arms," which prompted one observer to describe it as a "cross between a walrus and a giant squid." By 13 January, experts from the CSIRO had branded the carcass a slab of whale blubber. Ben Roesch readily agreed, but five years later, authors Loren Coleman and Patrick Huyghe found it "suggestive perhaps of a Giant Octopus."[252]

Greatstone-on-Sea, England: 1998

On 14 April 1998, young beachcombers Peter Jennings and Neil Savage found a peculiar eight-foot carcass beached at Greatstone, a village on the English Channel near New Romney in England's county of Kent. Fortean investigator Paul Harris alerted Dr. Karl Shuker to the stranding and furnished him with photographs of the carcass, which led to Shuker's identification of the globster as a rotting shark.[253]

St. Bernard's-Jacques Fontaine, Canada: 2001

Located on the east side of Fortune Bay, in the Canadian province of Newfoundland and Labrador, the separate fishing communities of St. Bernard's and Jacques Fontaine were legally amalgamated in 1994. Seven years later, on 2 August 2001, a globster beached nearby. It measured eighteen feet two inches long and resembled various others stranded on North Atlantic shores. A report published in the *Biological Bulletin*, in February 2002, identified the carcass as a decomposing sperm whale.[254]

Los Muermos, Chile: 2003

The new century's most famous globster - so far, at least - appeared in late June of 2003, on Pinuno Beach in the Chilean commune of Los Muermos, Llanquihue Province, in the nation's Los Lagos Region. The largest carcass washed ashore in modern times, it measured forty feet long and weighed an estimated thirteen tons. Elsa Cabrera, a marine biologist and director of the Centre for Cetacean Conservation in Santiago, told Reuters, "We don't know if it might be a giant squid that is missing some of its parts or maybe it's a new species. Apparently, it is a

The "Monster of Los Muermos" - a decomposing whale.

gigantic octopus or squid but that's just our initial idea, nothing definite. It has only one tentacle left. It could be a new species." [255]

James Mead, a zoologist with the Smithsonian Institution, took a more cautious approach. "I don't have enough data to say it's an octopus or its a whale," he said, "but I would hazard a bet that when it gets firmly identified, it'll be a whale. We're not going to know for sure on this specimen until someone gets a biological sample back to the laboratory." [256]

Another year elapsed before the final verdict was delivered, proving Mead's initial judgment correct. In June 2004, Sidney Pierce and five colleagues rendered their decision on the Chilean carcass (while inaccurately stating that it had been found on 26 July 2003, more than three weeks after BBC News aired the first report of its discovery). According to Pierce and company, molecular analysis of the globster "provide[d] irrefutable proof that the Chilean carcass was the highly decomposed remains of a sperm whale." Author Richard Ellis subsequently told the *New York Times,* "I'm crushed. It's a blow for people who continue to want there to be great and scary monsters out there." Still, he added, "It may be the requiem for blobdom, but there are other possibilities. We have yet to see a living adult representative of our friend the giant squid, so there's hope for monster watchers."[257]

Parton, England: 2004

Parton is a village on the coast of Cumbria in North West England, overlooking Solway Firth. In early September 2004, Joan Singleton found the remains of a "mini Loch Ness monster" beached nearby. A photo of the carcass, printed in the local *Whitehaven News* on 9 September, prompted fishing charter captain John Southam to tell the newspaper, "I have never seen anything like this in all my years at sea." Another local, unidentified, said, "It seems to have a seal's body, the tail of a whale, fins on top and sides, but also claws and really sharp teeth." Nor was the carcass a one-off, apparently. Whitehaven resident Robert Beattie claimed that he had seen a similar relic beached near Braystones, Cumbria, several weeks earlier. "It had quite big ribs and a backbone and I knew it could not be a fish, I thought it was an alligator, maybe a pet that had been thrown into the sea."[258]

Other pundits speculated that the carcass belonged to a seal pup or an "old penguin that has lost all its fur," while an anonymous caller told the *News* that he had spoken to "someone called CFZ" at the Centre for Fortean Zoology, evoking an urgent plea to retrieve the carcass. While no remains were forthcoming, the actual CFZ determined from published photos that the creature was a decomposed dolphin fetus. Local resident Frank Hewer agreed, writing to the newspaper on 24 September that:

> [w]ith regard to the dead sea mammal washed up on Parton beach, nowadays such an event is a rarity and seems to cause quite a stir. A dead porpoise (sea pig being the local name) washed up on Parton beach was not an uncommon sight when we were kids, as no doubt many others will well remember. [259]

A dissenting vote came from Dr. Brian McCusker at Newcastle University, who reported spotting "weird inbred beasts" four or five feet long, "like something from the *Alien* film," on fishing excursions to St. Bees and Seascale, Cumbria. In the absence of preserved remains, no definitive answer is available.[260]

Nea Kydonia, Crete: 2007

On 27 February 2007, one Regina Schmid uploaded a mystifying photograph to the American Greetings Webshots website. Viewable online at http://travel.webshots.com/photo/2657682110060412234hSUILr, it bears the caption "Carcass of unidentified animal on the beach Nea Kydonia," a municipality near Chania on the northwestern coast of Crete. Despite its intriguing label, the photo seems to show only a rocky beach strewn with logs, weeds and rubbish. The nearest thing to a carcass, size indeterminate, is a rounded, mottled object that might as easily be a stone or bit of driftwood. Since Ms. Schmid made no attempt to clarify the matter, it remains a mystery.[261]

Republic of Guinea: 2007

Three months after Regina Schmid posted her enigmatic photo from Crete (or wherever it was), Russia's *Pravda* website published five striking photos of a huge globster beached somewhere along the coast of Guinea, West Africa. Amazingly, the snapshots ran without a story to accompany them, providing no clue to precisely where the carcass came ashore or whether any part of it had been preserved for study.[262]

We should note here that while *Pravda* ("Truth") served as the Soviet Union's official propaganda voice from 1918 until 1991, today it is the equivalent of America's *Weekly World News,* replete with tales of alien invasions and female Yetis seducing "hot-blooded Caucasian men."[263]

Two views of a carcass beached on the coast of Guinea in 2007.

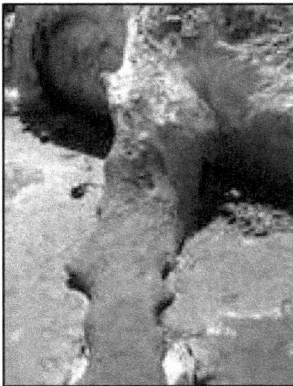

LEFT: An apparent tentacle extending from the Guinea carcass.

That said, the widely-published photos from Guinea prove that something was stranded in May 2007, and its size suggests that reference to rotted basking sharks must be mistaken.

Montauk, New York: 2008

Montauk is a hamlet on the southeastern shore of Long Island, New York, best known for its eighteenth-century lighthouse and as home to fictional shark-hunter Sam Quint in *Jaws.* On 12 July 2008, Jenna Hewitt and three friends reportedly discovered a small, peculiar carcass two miles from town, at a spot known as Ditch Plains Beach. Eleven days later, a local newspaper - *The Independent* - published Hewitt's photo of the creature and quoted her account of its discovery. "We were looking for a place to sit when we saw some people looking at something," she said. Upon seeing it, Hewitt recalled, "We joked that maybe it was something from Plum Island," home of an animal disease research facility operated by the Department of Homeland Security's Directorate for Science and Technology.[264]

Beyond its moment of discovery, mystery surrounds the "Montauk Monster." It was not preserved, nor even accurately measured, though one witness compared its size to that of a house cat. Its beak-like snout was puzzling, but visible ears and teeth, coupled with the lack of a shell or scales, torpedoed suggestions that it might be a decomposed turtle. Various published reports claimed that an unidentified man or women had removed the carcass to some undisclosed location and refused to share its whereabouts, for reasons still unclear.[265]

What was the creature? East Hampton Natural Resources Director Larry Penny opined that the carcass belonged to a raccoon with its upper jaw missing, a motion seconded by author Darren Naish. Internet blogger Hamilton Nolan found similarities between the Montauk carcass and X-rays of a water rat. William Wise, director of the Living Marine Resources Institute at New York's Stony Brook University, took a different view, describing the carcass as a fake created by "someone who got very creative with latex."[266] Without at least a fragment of the creature's flesh - or rubber, as the case may be - no definitive answer is possible.

Nunivak Island, Alaska: 2008

Ten days after the Montauk Monster surfaced, on 22 July, a larger carcass washed ashore on Nunivak Island, second-largest island in the Bering Sea, located thirty miles offshore from the delta of southwestern Alaska's Kuskokwim and Yukon Rivers. A photo of the pinkish corpse, apparently possessed of a long neck and tail, found its way onto Loren Coleman's Cryptomundo blog on 8 August. Three weeks later, Coleman wrote, "The whereabouts of the carcass, whether any samples were taken, or even if it has been revisited and bones gathered, are all unknowns."[267]

In fact, however, journalist Alex DeMarban reported on 7 August 2008 that samples of the creature's flesh *were* collected for study. Results of any tests performed remain unknown, but DeMarban quoted Mike Castellini, director of the Coastal Marine Institute at the University of Alaska Fairbanks, as saying that a photo of the beast was "sent to stranding experts and scientists as far as the Smithsonian and everyone is going, 'No idea.'" Castellini personally thought the carcass belonged to a mangled beluga whale, with key parts of its anatomy "obscured in the photo."[268]

A contrary view came from witness Barry Whitman, who snapped the original photo and sent it to Castellini. "I've seen decomposed walrus and whales and this was just something else," Whitman told DeMarban.

The creature's body measured six feet long, its tail three feet, and he suspected that its head was buried in the sand but made no effort to unearth it. Whitman's wife recalled that the tail terminated in a diamond-shaped appendage, leaving her "pretty baffled."[269] So we must remain today, with no lab results yet forthcoming.

New London, Connecticut: 2008

On 4 October 2008, Loren Coleman's blog ran photos of another stranded carcass, found by Bobette Clapsadle and her daughter at New London, Connecticut. Though visibly larger - or, at least, more bloated - than the Montauk Monster, its skull with teeth of an apparent carnivore intact resembled that of New York's specimen. Again, the creature seemed to be a quadruped and certainly a mammal, though some ignorant observers once again supposed it was a rotting turtle. Again, nothing suggests that any samples of the carcass were preserved for study, but an "official" Montauk Monster website noted archly that the new stranding site was "much closer to Plum Island than Montauk."[270]

Croyde Beach, England: 2009

On 6 January 2009, surfer Jason Poulton found a rotting carcass beached near Downend Point on Croyde Beach, in North Devon. It was "the size of a calf," he reported, "with massive canine teeth." Although badly decomposed, the five-foot-long body featured a coat of black fur that removed it from sea-serpent territory, while raising speculation about a very different cryptid. Specifically, the stranding occurred within five miles of the range where an elusive "black panther," dubbed the Beast of Exmoor, has been seen repeatedly since the 1970s. Occurring as it did on the same day that England's Forestry Commission officially confirmed the existence of exotic big cats roaming at large in Britain, the stranding fueled speculation that Exmoor's cat - or one of them, at least - had drowned and decomposed at sea. Thus far, as in the Montauk case and others, no report on any testing of the creature's flesh has been released. That paucity of data notwithstanding, Darren Naish concludes from photos that "the carcass is definitely that of a Grey seal" (*Halichoerus grypus*). [271]

Oxwich Beach, Wales: 2009

On 4 August 2009, hundreds of persons flocked to Oxwich Beach near Swansea, on the Gower Peninsula in Glamorgan County, Wales, to view a stranded creature described as "alien like" and resembling "something out of Dr. Who." London's *Daily Mail* offered two different descriptions of the beast within as many paragraphs on 5 August, first stating that the "writhing mass of tentacles" measured "at least 6ft from end to end," then contradicted that report by claiming that "the creature measured 3ft long." Whichever was correct, Professor Paul Brain of Swansea University soon recognized the object as a free-floating mass of goose barnacles (order *Pedunculata*).[272]

Bay of Islands, Newfoundland and Labrador: 2010

Newfoundland's Bay of Islands is a sub-basin of the Gulf of St. Lawrence, named for the myriad small islands scattered across it. On 24 February 2010, beachcomber Warrick Lovell found a globster somewhere on the shoreline, described as headless and fifteen feet long, with a "squarish torso," a ten-foot serpentine tail, and "what appears to be a single flipper-like appendage on its right side." Lovell quickly discounted his first impression of the carcass as a decomposing seal, and witness Rich Park abandoned his suspicion that it was a severed *Architeuthis* tentacle when he beheld the object's "hairy protrusion." An unnamed representative of the Department of Fisheries and Oceans came from Corner Brook, collecting samples of the creature's flesh for laboratory analysis, but no follow-up reports have been published thus far.[273]

Kitchenuhmaykoosib, Ontario: 2010

Three months after the Newfoundland stranding, another Canadian carcass surfaced at Kitchenuhmaykoosib, near northern Ontario's Big Trout Lake. Two tourists were hiking along a woodland creek, on 8 May 2010, when their dog called attention to the twelve-inch carcass. They photographed it, then left the area, later discussing their discovery with locals. Members of the Kitchenuhmaykoosib Inninuwug First Nation suggested that the beast might be a freshwater cryptid called *Omajinaakoo* ("the ugly one") in their native language, but when the two intrigued visitors tried to retrieve it, days later, no trace of the relic remained. While positive identification is therefore impossible, some observers of the photographs thought the mini-monster was a decomposed sea otter. Wilder opined that it might be the Puerto Rican *chupacabra,* several thousand miles away from home. Professional debunker Benjamin Radford examined a the photos and branded the carcass a decomposed "stinky mink."[274]

Tory Island, Ireland: 2010

Tory Island lies nine miles offshore from County Donegal, Ireland, in the North Atlantic. In June 2010, while visiting the island (population nineety-six that year), Yvonne Bolte found a desiccated carcass lying 330 feet inland and 165 feet above sea level. Its body had dried in a rough horseshoe curve, with a finned tail at one end and a wizened "face" at the other. Bolte wondered whether it was a mammal or reptile, but publication of her photo eleven months later, in *Fortean Times,* raised another possibility. In fact, the object bears a strong resemblance to a dried-out skate (family *Rajidae*), not unlike some of the home-made "Jenny Hanivers" offered as mermaids at various fairs and sideshows of yesteryear. In this case, no deliberate alteration of the carcass is apparent, though drying on shore has much the same effect.[275]

Milford, Connecticut: 2010

Witness Linda Ingmanson and her husband found our next globster in late September or early October 2010, reporting their discovery to the Montauk Monster website on 3 October. Their photos document the bloated, decomposing carcass of a quadruped, size indeterminate, its skull and feet denuded of flesh. Its tail was roughly equal in length to the creature's hind legs, and its skull appears to have lost all its teeth, though their sockets are clearly visible. Bloggers declared it "a mutated creature that bared resemblance to the Montauk Monster of recent years [sic]," but it could likely be identified by a mammalogist from photos of the skull, if any cared to try.[276]

Diggers Beach, New South Wales: 2010

Our last reported carcass - at least, for the present - came ashore at Diggers Beach, near Coffs Harbour, on the northern coast of New South Wales. Peter Atkinson found it on Australia's Father's Day, 5 September, and snapped a photograph of the two-foot-long creature before reporting it to the National Parks and Wildlife Department. His foresight was providential, since tides reclaimed the body prior to any official's arrival - and thus ruled out collection of the carcass for analysis.[277]

Facetiously dubbed a "sea monkey," the animal did resemble some kind of drowned primate, including small "hands" and a bushy tail. Its face, although averted from the camera, is obviously bare, as are its visible extremities; the rest, including its tail, is covered in thick grayish hair. Aside from the primate theory, some observers thought the animal might be a South American sloth, transported to Australia by means unknown. Lawrence Orel, speaking for National Parks and Wildlife, suggested that it might be a common brushtail possum (*Trichosurus vulpecula*), although the specimen lacked any trace of that semi-arboreal marsupial's prominent ears.[278]

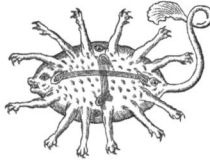

5.
Found Alive?

No discussion of globsters is complete without a review of cases wherein supposed sea or lake monsters have been captured or killed by intrepid adventurers. No such case has yet produced evaluation of a specimen by qualified professionals, but the reports are entertaining - and perhaps instructive - in their own right. The following thirty-six cases, then, are offered for whatever they are worth.

Gresskardfossen, Norway: ?

Peter Costello provides a list of nine captures or strandings in his classic 1974 work, *In Search of Lake Monsters*. The second on his list, immediately after Lake Mjøsa (see Chapter 1), indicates a case from Gresskardfossen, in Norway's Rogaland county. He provides no date, although the incident (and one other, below) appear on the chronological list prior to 1674. For whatever reason, Costello omitted the case from his text entirely - and, stranger still, I was unable to locate "Gresskardfossen" during my research for *Globsters*. It is not one of Rogaland's twenty-six present-day municipalities, and failed to surface during Internet searches.[1]

Reinsvatnet, Norway: ?

The second undated incident from Costello's roster, likewise omitted from his text, reportedly occurred at another Rogaland lake, called Reinsvatnet. He locates the incident at "Klype" - perhaps meaning the coastal municipality of Klepp - and claims that the creature was buried under a cairn which stands to this day. Chronologically, he places the event after the Gresskardfossen case and before 1674. Thankfully, we have a bit more information on this incident. Norwegian folklorist Torkell Mauland (1848-1923) alluded to an unknown creature being shot and killed after it was discovered stuck in a narrow pass. Its slayers then planted it under a cairn as described by Costello - but again, no dates or other details are provided. If the cairn indeed still stands, it might be worthy of investigation.[2]

Richelieu River, Québec: 1672

Records from the colonial era claim that a freshwater "siren" or monster was killed somewhere along the 106-mile course of this river, between Lake Champlain and Sorel-Tracy, Québec, at some time during 1672. No other details are available, but George Eberhart suggests that it was probably a seal.[3]

Lough Mask, Ireland: 1674

Lough Mask is a limestone lake spanning 22,000 acres in western Ireland's County Mayo. It is linked to Lough Corrib, farther south, which in turn is connected by the Corrib River to the sea, at Galway Bay. In 1684, author Roderick O'Flaherty told the following story concerning the death of a water monster at Lough Mask.

There is one rarity more, which we may term the Irish crocodile, whereof one, as yet living, about ten years ago had sad experience. The man was passing the shore just by the waterside, and spyed far off the head of a beast swimming, which he took to be an otter, and took no more notice of it; but the beast it seems lifted up his head, to discern whereabouts the man was; then diving swam under the water till he struck ground; whereupon he run out of the water suddenly and took the man by the elbow whereby the man stooped down, and the beast fastened his teeth in his pate, and dragged him into the water; where the man took hold of a stone by chance in his way, and calling to mind he had a knife in his jacket, took it out and gave a thrust to the beast, which thereupon got away from him into the lake. The water about him was all bloody, whether from the beast's blood, or his own, or from both he knows not. It was the pitch of an ordinary greyhound, of a black slimey skin, without hair as he imagines. Old men acquainted with the lake do tell there is such a beast in it, and that a stout fellow with a wolf dog along with him met the like there once; which after a long struggling went away in spite of the man and his dog, and was a long time after found rotten in a rocky cave of the lake when the waters decreased. The like they say is seen in other lakes in Ireland, they call it Dovarchu, i.e. water-dog, or anchu which is the same.[4]

Loch Leurbost, Lewis, Outer Hebrides: 1741/51

Loch Leurbost is one of many small lakes located in the southeastern sector of Lewis known as North Lochs (formerly the Parish of Lochs). Reports of a monster in the lake, published in March 1856, prompted correspondent "W.P." to send the following letter to the *Inverness Courier*:

> I beg to inform you that when shooting in that island in September 1821, with four gentlemen, we saw the same animal, and probably in the same loch; and for several hours endeavored to get an opportunity of shooting at the creature, but without success. We dined that evening with Mr. Mackenzie, at Stornoway, and mentioned what we had seen to him. Mr. Mackenzie expressed considerable surprise, but stated that the report was current in Lewis Island when he first came there, that such an animal had been captured in that very lake, and that it resembled in appearance a huge conger eel, and it required one of the farm carts to convey it to Stornoway. The capture of this creature must have occurred seventy or eighty years ago.[5]

Peter Costello, reporting that event, dated it "before 1776," but according to the letter we may peg it more precisely in the decade between 1741 and 1751.[6]

Norway: 1744

Erik Pontoppidan recorded our next case, describing it as follows:

> Anno 1744 one Dagfind Korsbeck catched, in the parish of Sundelvems on Sundmoer, a monstrous Fish, which many people saw at his house. It's [sic] head was like the head of a cat; it had four paws, and about the body was a hard shell like a Lobster's; it purred like a cat, and when they put a stick to it, it would snap at it. The peasants looked upon it as a Trold, or ominous Fish, and were afraid to keep it; and, consequently, a few hours later, they threw it into the sea again. According to the description, this might be called a Sea-Armadilla [sic], by which name an American Land-animal is known, nearly of the same shape, excepting that it has a long tail.[7]

"Sundmoer" today is the municipality of Sund, in Hordaland County, established in 1838. No size was offered for the beast, but Bernard Heuvelmans presumed that it must be fairly small, for Korsbeck to carry it home. Likewise, it cannot have been a known Atlantic fish, if it survived for hours out of water. Heuvelmans rightly deemed it "mysterious and difficult to place zoologically," finally treating it as a specimen of the cryptid he called a "super-otter."[8]

Atlantic Ocean: 1772

Our next case comes from Britain's *Annual Register* for 1772, rediscovered and sent to *Fortean Times* by Bristol correspondent Valerie Karatzas in 2004. It reads:

> *Paris, August 10.* Capt. Trebuchet, commander of a ship lately arrived in Nantes River, from St. Domingo, met with a very extraordinary event in his passage. The 16th day after he set sail, about 11 o'clock at night, he felt a great shock, and the whole crew imagined the ship had struck upon a rock; they immediately set the pumps to work, finding a great deal of water in the hold, and were all very much alarmed. When the day appeared, they found a monstrous fish, 30 or 40 feet long, fastened to the ship, and endeavoured by every means to get it off, but to no purpose.
>
> The captain therefore made up to a ship, about three leagues distant, which happened to be an English ship, commanded by Captain Smith, and with his assitance they at last cut away this monstrous fish; but it was then so much cut and disfigured that it was impossible to make out what it was, and they were afraid to send down divers to examine the damage done to the ship, for fear they should become prey to these voracious animals.
>
> The next day they examined the ship, and found her pierced in two

> places about four feet above her keel, by a kind of horn, which had made
> an orifice of three inches in diameter. They were obliged to pump night
> and day, and the English ship kept in company in order to give any
> assistance that might be necessary.

First, to clarify, there is no "Nantes River." Nantes is a French city on the Loire River, located thirty-one miles inland from the Atlantic coast. As to identifying the "fish," we may safely rule out swordfish (*Xiphias gladius*), sailfish (genus *Istiophorus*), marlin (genera *Makaira* and *Tetrapturus*), and sawfish (family *Pristidae*), since none of them attain the size described. Karatzas suggests an Arctic narwhal "swimming in the wrong ocean," but size remains a problem, since their record length is sixteen feet.[10]

Loblolly Cove, Massachusetts: 1817

From 1638 through the nineteenth century, New England residents were treated to successive visits by large marine cryptids. By 1817, monster-watching had became a significant spectator sport in Massachusetts, where several hundred witnesses observed huge sea-serpents cavorting offshore on various occasions. The existence of those great unknowns seems incontestable, yet they eluded stalwart fishermen and whalers, still remaining unidentified.

On 27 September of that year, it seemed that situation may have changed. A boy remembered only by his last name, Colbey, killed a most peculiar animal that afternoon, near Loblolly Cove, at Rockport, Massachusetts. The legless creature was three and a half feet long, with multiple humps on its back resembling those displayed by much larger serpents at sea. Young Colbey's father supposed it to be a baby sea-serpent and sold it as such to John Gott, at Sandy Bay. Gott, in turn, passed the carcass a Captain Beach Jr., who displayed it to the public at his home, then delivered it to the Linnaean Society of New England. That august body dissected the remains and reported that the deformed specimen "approaches most nearly to the *Coluber constrictor*" - a common nonvenomous snake, the eastern racer. Nonetheless, they proceeded to christen it *Scoliophis atlanticus* - the Atlantic humped snake - and thereby made themselves a laughingstock when the more prosaic identification was later confirmed.[11]

Raritan Bay, New Jersey: 1822

Bernard Heuvelmans recorded all that is known of our next captured monster, writing that :

> Charles-Alexandre Lesueur, the French zoologist, reported in 1822 that
> a huge sea-monster caught in Raritan Bay, New Jersey, was exhibited
> to the public as the 'Leviathan or Wonderful Sea-Serpent' although it
> was really a basking shark." He adds, furthermore, that it was "[p]
> robably the same monster, 30 feet long and about 18 feet in
> circumference, which according to the New York papers of 15 June
> 1822 was shot, harpooned and hauled ashore near Middleton Point. It
> had six rows of teeth and its liver yielded three barrels of oil.[12]

No "Middleton Point" exists in New Jersey, but three separate counties - Cape May, Monmouth and Morris - boast cities called Middle*town*. Raritan Bay is located in Raritan

County, adjacent to Monmouth.[13]

Chesapeake Bay, Maryland: 1840

Reports of a sea-serpent frequenting Chesapeake Bay - predictably nicknamed "Chessie" - have been logged for more than a century. Author Rosemary Guiley claims that one such was pursued, killed, and hauled ashore in August 1840. Furthermore, she says, "It was put on display at South Street wharf in Baltimore. It was described as being 12 feet long and nine feet [wide?] from fin to fin, and having a seal-like head and a large shell."[14]

Described by whom? Unfortunately, Guiley cites no source, and this sensational event slipped past Bernard Heuvelmans when he prepared his roster of sea-serpent cases in 1965. It also seems to have eluded every other author in the field, and proved untraceable at press time for *Globsters.*

Firth of Forth, Scotland: 1848

The Firth of Forth is the estuary of Scotland's River Forth, where it meets the North Sea between Fife and West Lothian. Dr. Heuvelmans reports that "[a]bout 1848 the fishing-smack *Sovereign* of Hull, fishing in the firth of Forth for Lord Norbury, caught a huge serpentine fish, 'which when spread out at length on the deck, extended beyond the limits of the vessel at stem and stern.' The fishermen were not surprised at its length, for they had come across an ever larger one, dark brown in color. The present fish was only from 4 to 9 inches thick with a dorsal fin 7 to 8 inches high. It was undoubtedly an oarfish, which in Scandinavia not even the dogs would eat, and Lord Norbury threw it overboard."[15]

Cullercoats, England: 1849

The next specimen offered for consideration appeared off Cullercoats, on the coast of North East England between Tynemouth and Whitley Bay. According to *The Zoologist,*

> *A strange marine animal*, of great size and strength, was captured on the 26th. of March off Cullercoats, near Newcastle. By the enclosed handbill, which hasbeen forwarded to me, it appears to be quite unknown to the neighbouring *savants*. The honest fishermen who drew the struggling monster to land are not, however, overscrupulous about the name, provided it be attractive enough to extract from the   pockets of 'ladies and gentlemen 6*d*.; working people 3*d*. each': they therefore boldly announce him as 'the great sea-serpent caught at last'. My correspondent very judiciously observes, that whatever the animal may be, it adds another to the many evidences constantly occurring that there *are* more things in heaven and earth, than are dreamt of by the most experienced practical observers. Some thirty five years since, the distinguished anatomist Dr. Barclay, was fain to reproach his contemporaries with the folly of affecting to suppose that they knew every thing. What additions have five and thirty years not given to Science! As the animal in question must be at least a local visitor, may we not hope, that some resident naturalist will favour us with a notice of it?[16]

Soon afterward, an announcement of the creature's public exhibition appeared, reading: "The great Sea-Serpent caught at last, by fourteen fishermen, off Cullercoats, on Monday last, March 26, 1849. This most wonderful monster of the deep was discovered by a crew of fishermen, about six miles from the land, who, after a severe struggle, succeeded in capturing this, the most wonderful production of the mighty deep. This monster has been visited by numbers of the gentry and scientific men of Newcastle, and all declare that nothing hitherto discovered in Natural History affords any resemblance to this. As an object of scientific inquiry,

this 'great unknown' must prove a subject of peculiar interest. Many surmises as to its habits, native shores, etc., have already been made, but nothing is really known. The general opinion expressed by those that are best able to judge, is, that this is the great sea-serpent, which hitherto has only been believed to have a fabulous existence, but which recent voyagers declare they have seen. Now exhibiting, at the shop, 57, Grey Street, opposite the High Bridge. Admission: ladies and gentlemen 6*d*., Working people 3*d*. each."[17]

The story began to change on 19 May, when the *Illustrated London News* reported:

> We observe in the Newcastle papers that a strange and hitherto unknown fish, nearly 13 feet in length, and possessing many of the characteristics which the captain of the *Daedalus* enumerated in his description of the great Sea-Snake, has really been caught off the Northumbriam coast, by the Cullercoats' fishermen, and has been exhibited in Newcastle, where it has created the greatest sensation. The members of the National History Society of that town have duly reported upon it, and expressed their opinion, that it is a young specimen of the genus *Gymnetrus*, only four of which species, and those very rare, are known to ichthyologists, and described by Cuvier and others as inhabiting the Indian, Mediterranean and White Seas. The present specimen has become the property of a Newcastle merchant, who has presented it to the museum of that town; and we understand that, in accordance with a very general wish of most of our distinguished naturalists, it is now exhibiting in the metropolis.[18]

Gymnetrus was the former name of what we now know as the oarfish, family *Regalecidae*. Today, all sources agree with the assessment published in May 1849.

Usan, Scotland: 1849

Our next case also comes to us from *The Zoologist*, regrettably without a date. Quoting a Scottish newspaper, the *Montrose Standard*, the journal offered this report:

> *A young sea-serpent.* - On Friday, while some fishermen belonging to Usan were at the out-sea fishing, they drew up what appeared to them a young sea-serpent, and lost no time in bringing the young monster to the secretary of our Museum. The animal, whatever it may be called, is still alive, and we have just been favoured with a sight of it; but

whether it really be a young sea-serpent or not, we shall leave those who are better acquainted with Zoology than we are to determine. Be it what it may, it is a living creature, more than 20 feet in length, less than an inch in circumference, and of a dark brown chocolate colour. When at rest its body is round; but when it is handled it contracts upon itself, and assumes a flattish form. When not disturbed its motions are slow; but when taken out of the water and extended, it contracts like what a long cord of caoutchouc would do, and folds itself up in spiral form, and soon begins to secrete a whitish mucous from the skin, which cements the folds together, as for the purpose of binding the creature into the least possible dimensions.[19]

A correspondent, one E. Newman, opined that the animal was a specimen of *Gordius marinus*, thereby creating some confusion. Historically, the name has been applied to two different marine worms. One, *Gordius marinus* Montagu, averages half an inch in length at maturity, while the other - now called *Lineus longissimus* - is one of the longest known living animals, with a specimen from 1864 exceeding 155 feet.[20]

Central Equatorial Pacific: 1852

Our next report, if true, ranks among the most dramatic in the annals of cryptozoology. Dr. Heuvelmans and Ben Roesch both dismissed it as a hoax, while authors Loren Coleman and Patrick Huyghe reserved judgment, noting simply that "others are not so sure." Karl Shuker also keeps an open mind, while leaning toward the skeptical.[21]

The facts - if facts they be - are these: the Massachusetts whaler *Monongahela* was searching for prey in latitude 3° 10'S and longitude 131° 50'W, when its lookout spotted a huge sea monster on 13 January 1852. Captain Charles Seabury, commanding the vessel, gave orders to pursue and kill the beast. According to a 2,700-word letter he wrote subsequently, describing the hunt, their prey was longer than his own 100-foot ship and measured nearly fifty feet in diameter. Its head was ten feet long and resembled an alligator's, with a mouth full of three-inch teeth. When killed and flensed, its body-fat yielded oil "clear as water" that "burnt nearly as fast as spirits of turpentine." Captain Seabury preserved the monster's head and heart for future study, before writing his report. On 6 February, after sailing around the Cape of Good Hope, the *Monongahela* met the brig *Gipsy*, bound for Bridgeport, Connecticut, and entrusted his letter to her commander, a Captain Sturges. Why he should do so remains unclear, but in fact the letter reached New England, while the *Monongahela* vanished forever at sea. Seabury's letter appeared in *The Times* of London on 10 March 1852.[22]

A century later, the story began to break down. Fortean author Frank Edwards, in his book *Stranger Than Science* (1959), revealed that there was no *Gipsy* or Captain Sturges. In fact, Seabury's letter was delivered by Captain Gavitt aboard the *Rebecca Sims*. That said, the *Monongahela* did exist, but her captain was *Jason* Seabury, not Charles. Photos of the ship existed, as of 1960, but the vessel and crew were lost. Tim Dinsdale reports that its name-board was found beached on Umnak Island, in the Aleutians, years after the supposed sea-serpent hunt.[23] It seems safest to dismiss the story as a hoax, and yet ....

Janesville, Wisconsin: 1853

In his book *Unnatural Phenomena*, Jerome Clark produced the following news item from the *Janesville Free Press*, reprinted in the *Alton* (Illinois) *Weekly Courier* of 18 March 1853. The date of the original remains unclear. It read:

> A singular reptile or fish was caught here a few days since, and is now in a glass jar before us. It has a skin like a catfish, a head and tail like an eel. The gills are on the outside of the neck, and it has four legs like a lizard, terminating in a miniature human hand. It is about fifteen inches long and was taken with a hook. No one here has ever seen a creature like it, though Thompson in his book of Vermont, describes a similar one, caught at Colchester, near Burlington.[24]

"Thompson," in this context, can only be famed Vermont naturalist Zadock Thompson (1796-1856), who indeed wrote several books about his native state's wildlife and wilderness. Without consulting them, it seems safe to say that the Janesville creature was a common mudpuppy or waterdog (*Necturus maculosus*), a salamander known to inhabit Wisconsin, which maintains its gills into adulthood and may reach a length exceeding nineteen inches.[25]

Bermuda: 1860

In 1860 *The Zoologist* offered the following item (undated, although independent reports say that it occurred on 22 January):

> *A sea-serpent in the Bermudas.* - I beg to send you the following account of a strange sea-monster captured on these shores, the animal being, in fact, no less than the great sea-serpent which was described as having been seen by Captain M'Quhae, of H. M. S. 'Daedalus', a few years since. Two gentlemen named Trimingham were walking along the shore of Hungary Bay, in Hamilton Island, on Sunday last, about eleven o'clock, when they were attracted by a loud rushing noise in the water, and, on reaching the spot, they found a huge sea-monster, which had thrown itself on the low rocks, and was dying from exhaustion in its efforts to regain the water. They attacked it with large forks which were lying near at hand for gathering in sea-weed, and unfortunately mauled it much, but secured it. The reptile was sixteen feet seven inches in length, tapering from head to tail like a snake, the body being a flattish oval shape, the greatest depth at about a third of its length from the head, being eleven inches. The colour was bright and silvery; the skin destitute of scales but rough and warty; the head in shape not unlike that of a bull-dog, but it is destitute of teeth; the eyes were large, flat, and extremely brilliant, it had small pectoral fins, and minute ventral fins, and large gills. There were a series of fins running along the back, composed of short, slender rays, united by a transparent membrane, at the interval of something less than an inch from each other. The creature had no bone, but a cartilage running through the body. Across the body at certain intervals were bands, where the skin was of a more

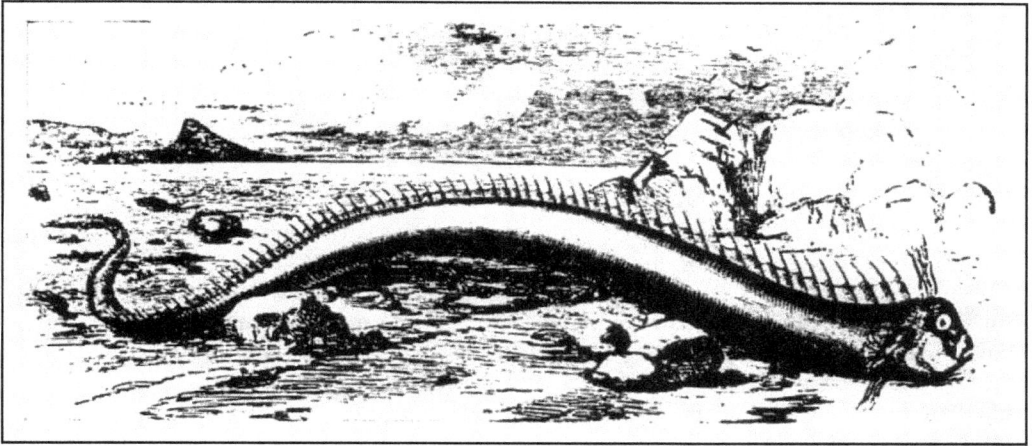

The Hungary Bay oarfish, 1860.

flexible nature, evidently intended for the creature's locomotion, screw like, through the water. But its most remarkable feature was a series of eight long thin spines of a bright red colour springing from the top of the head and following each other at an interval of about an inch; the longest was in the centre: it is now in the possession of Colonel Munro, the acting Governor of the Colony; and I had the opportunity of examining it very closely. It is two feet seven inches long, about three eighth of an inch in circumference at the base, and gradually tapering, but flattened at the extreme end, like the blade of an oar. The shell of these spines is hard, and, on examination by a powerful glass, appeared to be double, some red colouring matter being between the shells; the outside, which to the touch and natural eye was smooth, being rough and much similar to the small claws or feelers of the lobster or crayfish. The centre was a wide pith, like an ordinary quill. The three foremost of these spines were connected for about half their length by a greasy filament; the rest being unconnected; the serpent had the power of elevating or depressing the crest at pleasure. The serpent was carefully examined by several medical and scientific gentlemen; the head, dorsal spine, and greater part of the crest are in the possession of J. M. Jones Esq., an eminent naturalist, who will, doubtless, send home a more learned description of this 'wonder of the deep'. I regret that the immediate departure of the mail for England prevents my preparing you any more careful drawing of this great 'sea-serpent' than that I enclose.[26]

There can be little doubt that Anton Oudemans and Dr. Heuvelmans were correct in naming the Bermuda creature as yet another oarfish.[27]

North Sea: ca. 1875

Yet another oarfish was captured off the coast of Northumberland, England, on a date that Dr. Heuvelmans simply records as "before 1876." The animal was thirteen feet six inches long, and fifteen inches "deep." Based on his expertise, and without contrary evidence, we have no grounds for disputing his judgment.[28]

Bear Lake, Idaho: 1876

Bear Lake straddles the border of Utah and Idaho, with its 18.3-mile length almost equally divided on an east-west axis. In his 1974 treatise on lake monsters, Peter Costello reports that a twenty-foot cryptid was caught alive at Fish Haven in July 1876. Aside from misplacing the small community in Utah - it is actually on the Idaho side of the border - Costello quotes an unidentified newspaper as stating that the creature had a large mouth and "propelled itself through the water by the action of its tail and legs." Researchers Jerome Clark and George Eberhart found no trace of the incident while cataloging old cryptid reports, and since no follow-up stories appeared, Costello may be correct in his observation that the story was "perhaps a hoax, as such a capture ought to have made more of a stir."[29]

Oban, Scotland: 1877

Our next item comes from the august *New York Times,* published on 12 May 1877 under the headline "Capture of the Sea-Serpent.; A Serpent-Fish Over 100 Feet Long; Great Excitement - How the Thing Was Secured." The story read:

> The Glasgow *News* of a recent day publishes a circumstantial narrative by a resident at Oban, from which, if it be true, it appears that the sea-serpent has at length been actually captures at that place. Under date of April 27 the correspondent writes: "A most extraordinary event has occurred here, which I give in detail, having been eye-witness to the whole affair. I allude to the stranding and capture of the veritable sea-serpent in front of the Caledonian Hotel, George street, Oban. About 4 o'clock yesterday an animal or fish, evidently of gigantic size, was seen sporting in the bay near Heathen Island. Its appearance evidently perplexed a large number of spectators assembled on the pier, and several telescopes were directed toward it. A careful look satisfied us that it was one of the serpent species, it carrying its head 25 feet above the water. A number of boats were soon launched and proceeded to the bay, the crews armed with such weapons as could be got handy. Under the direction of Malcolm Nicholson, our boatman, they headed the monster and some of the boats were within 30 yards of it when it suddenly sprang half-length out of the water and made for the open. A random fire from several volunteers with rifles seemed to have no effect on it. Under Mr. Nicholson's orders the boats now ranged across the entrance to the bay, and by the screams and shouts, turned the monster's course, and it headed directly for the breast-wall of the Great Western Hotel. One boat, containing Mr. Donald Campbell, the Fiscal, had a most narrow escape, the animal actually rubbing against it. Mr. Campbell and his brother jumped overboard, and were picked up

unhurt by Mr. John D. Hardie, saddler, in his small yacht, the Flying Scud. The animal seemed thoroughly frightened, and as the boats closed in the volunteers were unable to fire more, owing to the crowds gathered on the shore. At a little past 6 the monster took the ground on the beach in front of the Caledonian Hotel, in George street, and his proportions were now fully visible. In his frantic exertions, with his tail sweeping the beach, no one dared approach. The stones were flying in all directions; one seriously injuring a man called Baldy Barrow, and another breaking the window of the Commercial Bank. A party of volunteers, under Lieut. David Menzies, now assembled, and fired volley after volley into the neck, according to the directions of Dr. Campbell, who did not wish, for scientific reasons that the configuration of the head should be damaged. As there was a bright moon, this continued till nearly 10 o'clock, when Mr. Stevens, of the Commercial Bank, waded in and fixed a strong rope to the animal's head, and by the exertions of some 70 folk it was securely dragged above high-water mark. Its exact appearance as it lies on the beach is as follows: The extreme length is 101 feet, and the thickest part is about 25 feet from the head, which is 11 feet in circumference. At this part is fixed a pair of fins, which are 4 feet long by nearly 7 feet across at the sides. Further back is a long dorsal fin, extending for at least 12 or 13 feet, and 5 feet high in front, tapering to 1 foot. The tail of more of a flattened termination to the body proper than anything else. The eyes are very small in proportion and elongated, and gills of the length of 2½ feet behind. There are no external ears; and, as Dr. Campbell did not wish the animal handled till he communicated with some eminent scientific gentlemen, we could not ascertain if there were teeth or not. Great excitement is created, and the country people are flocking to view it. This morning, Mr. Duncan Clark, writer, formally took possession of the monster, in the rights of Mr. M'Fee, of Appin, and Mr. James Nicol, writer, in the name of the Crown. [30]

Two days later, the *Times* ran two follow-up stories. One discoursed at length on the alcoholism of Scotsmen, while the other ran under the headline "The Sea-Serpent Hoax." It read:

A London newspaper of May 2 says: "Mr. Wybrow Robertson, of the Westminster Aquarium, writes that, in consequence of the detailed statement published respecting the capture of the sea-serpent, he at once telegraphed to Mr. Duncan Clark, Writer, Oban, offering to purchase the creature for exhibition. In reply Mr. Robertson received the following telegram: 'The whole thing is a shameful hoax deserving no attention, except to punish the author.'"[31]

Tasmania: 1878

On 6 September 1878, the *Scotsman* published the following story under the headline "A Baby Sea-Serpent":

From Van Diemen's Land [Tasmania] comes news of the capture of a queer fish. It is fourteen feet long, fifteen inches deep from the neck to the belly, tapering two inches to the tail, and eight inches in diameter in the thickest place. There are no scales, but the skin is like polished silver, with eighteen dark lines and rows of spots running from the head to the tail each side. There is a mane on the neck twenty inches long, and continues from the head to the tail; small head, no teeth, protrusive mouth, capable of being extended four inches like a sucker; eyes flat, about the size of a half crown, and like silver, with black pupils. There are two feelers under the chin, thirty-two inches long. The fish was alive when captured.[32]

Six days later, *Nature* published a letter from one Andrew Wilson, suggesting that the story from Tasmania written "is explicable only on the tape fish theory," referring to an oarfish.[33] Based on the animal's description, he was undoubtedly correct.

New Harbor, Maine: 1880

Our next strange specimen, still hotly debated, was found floating dead off New Harbor, Maine, in August 1880. Discovered by Captain S.W. Hanna, the elongated carcass measured twenty-five feet long and nine inches in diameter at its thickest point. A sketch done at the time depicts a flat face with a slight protrusion over a small mouth, gill slits, small pectoral fins on the same axis as a triangular dorsal fin, and another fin that encircled the tip of its tail, resembling an eel's caudal fin. Hanna said that the creature's back was "slate or fish colored," while its belly was "grayish white." Irritated that his net had torn while hauling in the creature, Hanna tossed it overboard.[34]

What was it? Three authors - Dr. Heuvelmans, Dr. Shuker, and Michael Bright - all cast their votes for an unknown species of serpentine shark. Bright, in fact, proclaimed, "There is no doubt that it was a largish relative of *Chlamydoselachus*" - the frilled sharks. Two known species exist, and while they match Captain Hanna's sketch in some respects, both have small dorsal fins set far back on the body, away from the pectoral fins. In terms of size, the record specimen of *Chlamydoselachus africana* was under four feet long, while the largest representative of *C. anguineus* measured six feet six inches, just over one-fourth the length of Hanna's catch.[35]

Ben Roesch took a contrary view, writing that the animal in question "is more likely to be a

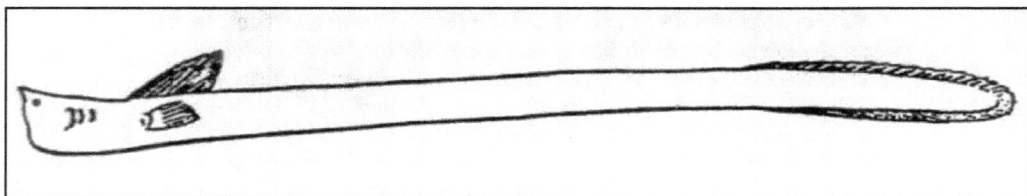

Captain S.W. Hanna's mystery fish, 1880.

bony fish than a shark, and that it seems it represents a new species of elongate bony fish, which remains undiscovered." While dodging a specific I.D., Roesch offered illustrations of two crestfish species (family *Lophotidae*) that he deemed comparable in form. Again, however, there are drawbacks: living crestfish are known only from deep tropical and subtropical waters (which should exclude New England), and the largest known species (*Lophotus capellei*) boasts a record length of six feet six inches.[36]

Haro Strait, British Columbia: 1880

Haro Strait lies between Vancouver Island and British Columbia's Gulf Islands, linking the Strait of Georgia to the Strait of Juan de Fuca. On 23 September 1880 the *New York Times* reported a strange catch from those waters, made on the previous day. That article read:

A dispatch from Victoria says:

> A genuine sea serpent, 6 feet in length, with an orthodox mane, head shaped like a panther's, and a tail whittled down to a sharp point, was brought in by Indians yesterday, they having caught it in deep water in the Straits of Deharo. Its appearance creates intense interest among savants, and old fishermen cannot place the monster. The serpent has been photographed, and the body will be preserved in spirits and sent to Ottawa for classification.[37]

There ends the tale, with no trace of the photos or the creature mentioned subsequently. Ben Roesch speculated that it may have been an oarfish or a ribbonfish (family *Trachipteridae*), but in the absence of that tantalizing pickled specimen, who knows?[38]

Marin County, California: 1883

Three years later and 800 miles farther south, the *New York Times* offered another sea-serpent story, this one reprinted from the *San Francisco Bulletin* on 15 September 1883. The original item appeared on 7 September and read:

Messrs. L. Laveosa & Co., at the California Market, think that they have a veritable sea-serpent, but it is not so large as the specimens seen at various watering-places just at the opening of the season. It is 8 feet long, 4½ inches in circumference, copper-colored on top with dark brown spots which extend from the head to the tail, and with a pale brown belly. It has no gills nor dorsal fin, and its tail tapers to a point like that of an ordinary snake. The mouth is 2 inches long, armed with two rows of sharp and retentive teeth, but devoid of fangs. It looks like a reptile, and proved to be so vicious when caught in a net by Italian fishermen off the Marin County shore last Tuesday [4 September] that they clubbed it into quiet. It was still alive and still ferocious Wednesday, which day it died. Naturalists may view it with interest but it would have been a better study alive.[39]

Indeed it would! Once more, we are confronted with the riddle of a supposed fish living for twenty-four hours or more out of water - and after a clubbing, at that. Dr. Heuvelmans suggested that it may have been a California moray eel (*Gymnothorax mordax*), and Ben

Roesch concurs. Various anecdotal reports describe morays surviving out of water for protracted periods, if their skins remain moist.[40]

Río Beni, Bolivia: 1883

The Río Beni flows through the La Paz Department of northern Bolivia. In 2002, Richard Eberhart cited a 119-year-old article from *Scientific American,* describing a "saurian" with a doglike head killed along the river by hunters. It measured thirty-six feet long, had scaly skin, and allegedly sported two smaller "heads" sprouting from its back. No further mention of the beast was made after its slayers sent the carcass to La Paz.[41]

The largest known reptile in Bolivia is the green anaconda (*Eunectes murinus*). Stories persist of specimens exceeding thirty feet, though none have been scientifically confirmed. In theory, a large anaconda afflicted with tumors might seem to have multiple heads if viewed by superstitious and uneducated hunters, but any attempt to identify the Río Beni creature remains speculative - assuming it ever existed at all.

North Atlantic: 1885

Jerome Clark found our next item in an issue of the *Elyria Weekly Republican,* published on 30 July 1885. Although printed in Ohio, it recounts supposed events from Porrtland, Maine, where a Captain Cobb of the schooner *Dreadnought* allegedly delivered a sea monster caught five miles off Halfway Rock, in Essex County, Massachusetts. Despite multiple harpoons piercing its flesh, the 1,200-pound creature - described as "a cross between a catfish and a snake," protected by "some kind of armor that may be called his shell" - remained lively on arrival at Portland's Custom House wharf. According to the article, "Dr. Thomas Hill is to be appealed to and asked to name the monster. In the mean time, the snake or turtle, as the case may be, promises to get well of his wounds and manifests a strong desire to smash things."[42] In the absence of confirmation, which was never forthcoming, we are safe in dismissing the tale as a hoax.

Ballynahinch, Northern Ireland: 1888

During 1888, the same year when a giant eel allegedly got stuck and died at Crolan Lough in Connemara (see Chapter 4), a similar monster was trapped beneath a bridge near Ballynahinch Castle, in Northern Ireland's County Down. Witnesses estimated the creature's length at thirty feet, but unlike its hapless relative in the south, it survived for several days and finally escaped when the river's level rose and washed it free. Third-hand accounts of the event surfaced eight decades later, and remain unverified.[43]

Jacksonville, Florida: 1890

Phenomenal researcher Jerome Clark unearthed another report from Ohio, pertaining to far-off events, in an issue of the *Marion Daily Star,* published on 13 May 1890. This article describes the capture of a mermaid by fisherman W.W. Stanton, aboard the schooner *Addie Schaeffer.* He supposedly delivered it alive to Jacksonville, where it survived for two days, uttering "a low, moaning sound, which might easily hae been mistaken for the sobbing of a baby." Following its death, the creature was reportedly preserved in a huge bottle of alcohol. According to the *Daily Star*'s account:

This strange creature is about six feet long, pure white and scaleless. The head and face are wonderfully human in shape and feature. The shoulders are well outlined, and very much resemble those of a woman, and the bosom is well defined and shows considerable development, while the hips and abdomen continue the human appearance. There are four flippers, two of which are placed at the lower termination of the body, and give the impression that nature made an effort to supply the creature with lower limbs.[44]

By now, it comes as no surprise that follow-up reports on this remarkable event are nonexistent. If the mermaid did exist, it somehow vanished from its bottle - and from history.

Atlanta, Georgia: 1891

Clark's next discovery, from May 1891, is both more plausible and has the added benefit of coming from a newspaper in the vicinity of the event. On 26 May, the *Atlanta Constitution* told its readers that "[a] large and curious fish was caught in the creek near here [Calhoun, Ga.] recently. It's [*sic*] head resembled that of a snake, and it had teeth like a human being. It is a variety unknown to the oldest fishermen."[45]

There are "snakehead" fish (family *Channidae*), native to Africa and Asia, which have found their way into North American waters, and have featured in several low-budget horror films during the twenty-first century. That said, their first confirmed sighting in the United States - at a pond in Crofton, Maryland - did not occur until 2002. Furthermore, the snakeheads' sharply-pointed teeth bear no resemblance whatsoever to a human's.[46] Whatever the Georgia specimen may have been, it remains unidentified today.

Gulf of Mexico: 1896

In August 1896 the *Shipping Gazette* published the following account from Carrabelle, Florida (Franlkin County):

A sea-serpent is reported to have been captured at Carabelle [*sic*], Florida, by a fishing vessel named the *Crescent City,* which it towed wildly for a time before it was killed.The thing measured 49 feet long and 6 feet in circumference. It is eel shaped, with a shark-like head and a tail armed with formidable fins. It was caught with a shark hook, but after being tired out it had to be shot.[47]

Dr. Heuvelmans initially considered the Carrabelle creature another specimen of eel-like shark, perhaps akin to Captain Hanna's catch from the North Atlantic, but his final roster lists it as a whale shark or basking shark, mitigated by a speculative question mark. No further information is available today.[48]

Mud Lake, Arkansas: 1897

Arkansas claims nineteen bodies of water called "Mud Lake," scattered over fourteen counties. Arkansas County boasts four of its own, while Monroe County has two. Two more reside in St. Francis County, home of the *Forrest City Times* which reported the following tale on 28 May 1897. Sadly, we can come no closer to pinpointing the event, if in fact it ever happened.[49]

As summarized by George Eberhart, the article claimed that "a 16-foot monster with scaly skin" had been harpooned by locals at Mud Lake. American alligators may exceed that length in rare cases, and might follow the Mississippi River northward from their normal range in southern bayous, but it seems unlikely that local hunters would mistake a common reptile for some unknown species. Perhaps the story was another hoax, as hinted by its headline: "Rather Fishy."[50]

Racine, Wisconsin: 1902

On 10 July 1902 the *Milwaukee Journal* reported that William Wuertzberger, of Racine, had "dipped a curious fish" from Lake Michigan. As described in the article:

It has the head of a lizard and body of a fish, is fourteen inches long and two-and-a-half inches in diameter, of grayish color and with black spots. It has four feet, resembling those of a lizard, but much smaller and the tail of an eel. When placed in the water with other fish it emitted pills which dissolved and killed the other fish. There were no eyes. There are two small ears, an eighth of an inch in diameter, but when the fish became angry would extend over an inch.[51]

This may have been another common mudpuppy. It falls within the known size range, is gray with occasional darker spots, and its fanlike gills might be mistaken for ears. Amphibian eggs resemble "pills" released into water, but their lethal effect on fish remains unexplained.

Copeland Islands: 1908

The Copelands are a cluster of three islands - Copeland, Mew and Lighthouse - lying in the Irish Sea off Donaghadee, Northern Ireland. On 11 September 1908, the weekly *County Down Spectator* reported an event occurring six days earlier. That article read:

> Following the reports circulated recently as to the present of a strange sea monster in Belfast Lough, a letter has been received from a resident of the Copeland Islands. He states that on the 5th last great excitement was caused on the islands when it became known that a huge snake-like fish had been stranded in the shoal on Horse Point.

This correspondent states that he and his brother were out walking, when they observed the water in the shoal being lashed about as if by a whale. The tide was out at the time, and on approaching the spot they were amazed to see a monster fish swimming about. Too terrified to get any closer they were at a loss what to do; but at length, the correspondent, realising that the incoming tide would liberate the monster, despatched his brother for a gun and told him to bring the boat round.

"It took us all our time," he states, "to kill the beast, and it was only after four shots had been fired into him that he stopped kicking. We then grappled him, but try as we might we could not get him to budge, so John went and brought two other men and a pony, and amongst us we beached him at last." Describing his capture, our correspondent states that he measured it and found it to be nearly

30 foot long, and about 6 feet round at its upper fins. The body tapers to about 6 inches at the tail, which is fan-like. There are three large fins, two on the back and one on the belly. The mouth, nose, and eyes resemble those of a conger eel, but are about five times as large. The body is covered in scales. The writer says that he is an old man, has lived all his life on the Copelands, and has seen most queer fishes, but never anything like this. He states that if any Belfast gentleman would care to examine the monster he or any of the residents of the island would on being signalled for take them from Donagharlee pier to where the body is beached. He adds that he would have communicated with us sooner but for the fact that during the past two days the weather has been too wild to permit of getting across the mainland.[52]

Based on the article's description, Ben Roesch suspected that the creature was an oarfish, though he granted that the placement of its fins was "somewhat strange." Likewise, description of the monster's scales would have to be erroneous, since oarfish have none. Despite those "minor discrepancies," Roesch stands by his judgment, and in the absence of substantive evidence no one can prove him wrong - or right.[53]

Galveston, Texas: 1910

On 21 August 1910 the *Galveston Daily News* reported that a "monster unknown by name in the fish world which was caught up on the jetties yesterday, will be on exhibition to-day at the Electric Park."[54] No follow-ups were run, and there the matter rests, without even a hint of the creature's description.

Venice, California: 1912

Jerome Clark found our next item in Oklahoma's *Ada Evening News,* inexplicably published 1,238 miles from the scene of the event on 29 November 1912. The article informs us that:

> One of the queerest deep sea creatures ever seen in this vicinity was brought in a few days ago by a fisherman of Venice, Cal. It is five feet in length, black and green mottled, with a tail like that of a shark. It has a dorsal fin and four feet, shaped like those of a parrot. Its mouth resembles that of a Gila monster, while its head is a replica on a large scale of that of a California horned toad.[55]

Perhaps. But in the absence of corroborating evidence, that hybrid monster seems too "queer" to be true.

DeCourcy Island, British Columbia: 1968

In their study of *Cadborosaurus*, authors Paul LeBlond and Edward Bousfield record the claim of retired whaler William Hagelund that he caught "a very young specimen of Caddy" while yachting through British Columbia's Gulf Islands, in August 1968. The event supposedly occurred at Pirate's Cove, on DeCourcy Island, when Hagelund and his sons spied "a small eel-like creature swimming along with its head completely out of the water, the undulation of its long, slender body causing portions of its spine to break the surface." It possessed "dark limpid eyes, large in proportion to the slender head, which had given it a seal-like appearance when viewed from the front."[56]

Hagelund says they caught the animal and kept it aboard their boat overnight, in a bucket of water. It was sixteen inches long and just over an inch in diameter, with tiny sharp teeth set in its lower jaw. The back was covered by "plate-like scales," while "soft yellow fuzz" coated the underside. Two small "flipper-like feet" protruded around shoulder-level, while the serpentine body ended in a "spade-shaped tail [that] proved to be two tiny flipper-like fins that overlapped each other." Fearing it would die in the bucket, Hagelund released the creature before daybreak - and neglected to snap any photos beforehand. LeBlond and Bousfield refrain from judging Hagelund's account, first aired in a memoir he published in 1987, and readers must take it or leave it, as they see fit.[57]

Kilwa Kisiwani, Tanzania: 1975

Kilwa Kisiwani is an island community located off the southern coast of Tanzania, East Africa. In May 1975 a local fisherman reportedly netted a "fish-like creature" unrecognized by island residents. That was not surprising, since reports broadcast from Dar-es-Salaam claimed the "fish" had arms and legs "with the human complement of toes," prominent eyes - one of them glowing - and a horn (presumably atop its head). The staff of *Fortean Times* alerted contacts at the British Natural History Museum, but no more was heard of the remarkable African catch.[58]

Johns Island, Washington: 1991

Retired Seattle pharmacist Phyllis Harsh claims that she found a living "baby dinosaur" on Johns Island, in the San Juan Islands archipelago, during July 1991. Using a pair of sticks, she helped the two-foot-long creature back into the surf, from which it disappeared into Haro Strait. Authors Paul LeBlond and Edwward Bousfield consider the specimen an infant *Cadborosaurus,* noting that Harsh also claimed discovery of a second "small dinosaur" - this one a skeleton, found in a bald eagle's nest - on Johns Island. She did not preserve the skeleton for study.[59]

Selangor River, Malaysia: 2006

The Selangor River is a major waterway in Selangor state, Peninsular Malaysia, flowing from Kuala Kubu Bharu to reach the Strait of Malacca at Kuala Selangor. On 25 January 2006, while searching for bait at the river's Telok Gong estuary, native fisherman Arbain Sellah saw monitor lizards feeding on a large, unrecognized object. A closer look revealed the decomposed remains of some creature, half-buried in sand. Some locals thought it might be the carcass of a dwarf crocodile (*Osteolaemus tetraspis*), but photos of the skull and skeleton bore no resemblance to a crocodilian's. Biologist Lateffah Sugito, from Terengganu's Turtle and Marine Ecosystem Center, opined that the bones belonged to a marine animal, and while she ventured a guess that it might be a dugong, no official confirmation of that guess has yet been published.[60]

Trinity Bay, Newfoundland and Labrador: 2009

John Marsh is a veteran fisherman, with six decades of experience operating from Lower Lance Cove, on Newfoundland's Trinity Bay. Still, he had never seen anything vaguely resembling the creature that found its way into one of his caplin traps during summer 2009.

"It's almost too strange to talk about," he told the St. John's *Telegraph* in March 2010. "It almost don't sound real, but I told you the story of it and we've seen it."[61]

"It" was large, with a sinuous neck eight to ten feet long, and blue-green skin "smooth as glass." While struggling to free the creature, Marsh noted its lack of a cetacean blowhole. Furthermore, he said, "If it was a whale or anything like that he would have had old barnacles and scratches on it and stuff like that, but this was perfectly clean just like he come from a washer." Caught without a camera, Marsh and his crew could only watch the creature swim away, once it was freed.[62]

Informed of the incident months later, Department of Fisheries and Oceans mammalogist Jack Lawson bemoaned March's failure to mutilate his healthy catch. "Obviously," Lawson told the *Telegraph,* "if he saw something, it would have been great if he could have cut a piece off it or kept it. That's the frustrating part for me. I would have loved to have seen what it was." Lawson surmised that Marsh may have seen *Cadborosaurus,* though that name properly applies to a cryptid reported from Pacific Northwest waters, some 3,100 miles to the west. In closing, Lawson told the newspaper, "I love a mystery!"[63]

So do we all. And even though analysis of our 132 globsters and thirty-six alleged captures finds most to be explainable as known species or hoaxes, the mystery endures.

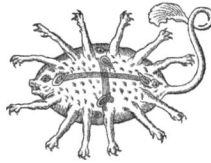

Notes

Chapter 1

1. Bryan Dunning, "Attack of the Globsters!" Skeptoid No. 152 (5 May 2009), http://skeptoid.com/episodes/4152.

2. Ellis, *Men & Whales,* p. 36; "Beached whale," Wikipedia, http://en.wikipedia.org/wiki/Beached_whale.

3. Carwardine, pp. 116-17; "Andrew's beaked whale," Wikipedia, http://en.wikipedia.org/wiki/Andrews%27_Beaked_Whale.

4. Carwardine, pp. 132-3; "True's beaked whale," Wikipedia, http://en.wikipedia.org/wiki/True%27s_Beaked_Whale.

5. Carwardine, pp. 134-5; Coleman and Huyghe, p. 286; "Tropical bottlenose whale," Wikipedia, http://en.wikipedia.org/wiki/Longman%27s_Beaked_Whale.

6. Carwardine, pp. 114-15; "Sowerby's beaked whale," Wikipedia, http://en.wikipedia.org/wiki/Sowerby's_Beaked_Whale.

7. Carwardine, pp. 122-3; Coleman and Huyghe, p. 286; "Gervais' beaked whale," Wikipedia, http://en.wikipedia.org/wiki/Gervais%27_Beaked_Whale.

8. Carwardine, pp. 124-5; "Ginkgo-toothed beaked whale," Wikipedia, http://en.wikipedia.org/wiki/Ginkgo-toothed_Beaked_Whale.

9. "Spade-toothed whale," Wikipedia, http://en.wikipedia.org/wiki/Spade_Toothed_Whale.

10. Carwardine, pp. 130-1; "Strap-toothed whale," Wikipedia, http://en.wikipedia.org/wiki/Strap-toothed_Whale.

11. Carwardine, pp. 128-9; "Hector's beaked whale," Wikipedia, http://en.wikipedia.org/wiki/Hector%27s_Beaked_Whale.

12. Carwardine, pp. 140-1; "Shepherd's beaked whale," Wikipedia, http://en.wikipedia.org/wiki/Shepherd%27s_Beaked_Whale.

13. Carwardine, pp. 138-9; "Stejneger's beaked whale," Wikipedia, http://en.wikipedia.org/wiki/Stejneger%27s_Beaked_Whale.

14. Carwardine, pp. 142-3; Coleman and Huyghe, p. 286; "Cuvier's beaked whale," Wikipedia, http://en.wikipedia.org/wiki/Cuvier's_Beaked_Whale.

15. Carwardine, pp. 136-7; "Pygmy beaked whale," Wikipedia, http://en.wikipedia.org/wiki/Pygmy_Beaked_Whale.

16. "Five basking sharks found dead on coast," BBC News, 18 June 2004.

17. "Whale shark," Wikipedia, http://en.wikipedia.org/wiki/Whale_shark.

18. "Basking shark," Wikipedia, http://en.wikipedia.org/wiki/Basking_shark.

19. "ISAF Statistics on Attacking Species of Shark," International Shark Attack File, http://www.flmnh.ufl.edu/fish/sharks/statistics/species2.htm.

20. "Great white shark," Wikipedia, http://en.wikipedia.org/wiki/Great_white_shark.

21. J.E. Randall, "Size of the great white shark (Carcharodon). Science 181 (1987): 169-70; Jennifer Viegas, "Largest Great White Shark Don't Outweigh Whales, but They Hold Their Own," Discovery Channel, http://dsc.discovery.com/sharks/largest-great-white-shark.html; S. Wroe, et al., "Three-dimensional computer analysis of white shark jaw mechanics: how hard can a great white bite?" *Journal of Zoology* 276 (December 2008): 336-42.

22. "Great hammerhead," Wikipedia, http://en.wikipedia.org/wiki/Great_hammerhead.

23. "Distribution Table of Confirmed Megamouth Shark Sightings," Florida Museum of Natural History, http://www.flmnh.ufl.edu/fish/sharks/megamouth/tablemega.htm.

24. "Tiger shark," Wikipedia, http://en.wikipedia.org/wiki/Tiger_shark; Summary of Large Tiger Sharks *Galeocerdo cuvier,* http://homepage.mac.com/mollet/Gc/Gc_large.html.

25. "Beached Cow Mistaken For Polar Bear," 21 September 2010, http://www.popfi.com/2010/09/21/beached-cow-mistaken-for-polar-bear.

26. "Treasure trove of new marine species found," MSNBC (19 September 2006), http://www.msnbc.msn.com/id/14834763.

27. Steve Connor, "Thousands of new marine species found in Pacific's 'golden triangle,'" The Independent (London), 6 February 2007.
28. "New Marine Species Discovered In Eastern Pacific," ScienceDaily (9 March 2007), http://www.sciencedaily.com/releases/2007/03/070308121755.htm.

29. Carolyn Barry, "Nearly 300 New Marine Species Found Near Australia," National Geographic News (9 October 2008), http://news.nationalgeographic.com/news/2008/10/081009-new-marine-life.html.

30. Adam Morton, "Rare prawn found amid 6000 new marine species," *The Age* (Melbourne, Australia), 5 October 2010.

Chapter 2

1. "Kraken," Wikipedia, http://en.wikipedia.org/wiki/Kraken.

2. Ibid.

3. Heuvelmans, *The Kraken and the Colossal Octopus,* pp. 157-64.

4. Ibid., pp. 155-7.

5. "List of giant squid specimens and sightings," Wikipedia, http://en.wikipedia.org/wiki/List_of_giant_squid_specimens_and_sightings.

6. Ibid.; Ellis, *The Search for the Giant Squid,* pp. 65-7.

7. "List of giant squid specimens and sightings"; Ellis, *Search,* pp. 67-8.

8. Ellis, *Search,* p. 72.

9. "List of giant squid specimens and sightings."

10. Ellis, *Search,* p. 73.

11. "Giant squid," Wikipedia, http://en.wikipedia.org/wiki/Giant_squid.

12. "List of giant squid specimens and sightings."

13. "Giant squid."

14. Steve O'Shea and Kat Bolstad, "Giant Squid and Colossal Squid Fact Sheet," The Octopus News Magazine Online, http://www.tonmo.com/science/public/giantsquidfacts.php.

15. "List of giant squid specimens and sightings"; Ellis, *Search,* pp. 205, 258-60; Heuvelmans, *Kraken,* pp. 238-9.

16. Ellis, *Search,* p. 106.

17. Ibid., pp. 203-5.

18. Ibid., pp. 245-7.

19. Heuvelmans, *Kraken,* pp. 224-5.

20. Ibid., p. 234.

21. Ibid., pp. 234-5.

22. Ibid., p. 236.

23. Ibid., p. 235.

24. Ibid., p. 235; Richard Young and Michael Veccione, *Sthenoteuthis oualaniensis,* Tree of Life Web Project, http://tolweb.org/Sthenoteuthis_oualaniensis.

25. Heuvelmans, *Kraken,* p. 235.

26. Ellis, *Search,* p. 147; "List of Colossal Squid specimens and sightings," Wikipedia, http://en.wikipedia.org/wiki/List_of_Colossal_Squid_specimens_and_sightings.

27. ""List of Colossal Squid specimens and sightings."

28. "Colossal Squid," Wikipedia, http://en.wikipedia.org/wiki/Colossal_Squid.

29. "Galiteuthis phyllura," Wikipedia, http://en.wikipedia.org/wiki/Galiteuthis_phyllura.

30. "Robust Clubhook Squid," Wikipedia, http://en.wikipedia.org/wiki/Moroteuthis_robusta.

31. "Megalocranchia fisheri," Wikipedia, http://en.wikipedia.org/wiki/Megalocranchia_fisheri.

32. "Humboldt Squid," Wikipedia, http://en.wikipedia.org/wiki/Dosidicus_gigas; Ellis, *Search,* p. 109.

33. "Giant Squid Found," *MonsterQuest,* History Channel, 14 November 2007; "Giant Squid Ambus," *MonsterQuest,* History Channel, 8 October 2008.

34. "Dana Octopus Squid," Wikipedia, http://en.wikipedia.org/wiki/Taningia_danae; Ellis, *Search,* pp. 149-52.

35. "Kondakovia longimana," Wikipedia, http://en.wikipedia.org/wiki/Kondakovia_longimana; Damian Carrington, "Big squid breaks record," BBC News, 3 July 2000.

36. "Diamond Squid," Wikipedia, http://en.wikipedia.org/wiki/Thysanoteuthis_rhombus.

37. Bijal Trivedi, "'Weird' New Squid Species Discovered in Deep Sea," National Geographic Today (20 December 2001), http://news.nationalgeographic.com/news/2001/12/1220_TVweirdsquid.html.
38. Ibid.

Chapter 3

1. "Giant octopus," Wikipedia, http://en.wikipedia.org/wiki/Giant_octopus.

2. "North Pacific Giant Octopus," Wikipedia, http://en.wikipedia.org/wiki/North_Pacific_Giant_Octopus; "*Enteroctopus dofleini*, Giant Octopus," Marinebio, http://marinebio.org/species.asp?id=60.

3. Frank Lane, *Kingdom of the Octopus* (New York: Pyramid, 1960), p. 22.

4. Mark Norman, *Cephalopods: A World Guide* (Heckenhaim, Germany: ConchBooks, 2003), pp. 213–216.

5. "Seven-arm Octopus," Wikipedia, http://en.wikipedia.org/wiki/Haliphron_atlanticus.

6. Heuvelmans, *Kraken,* p. 273; Mackal, *Searching,* pp. 36-7.

7. Heuvelmans, *Kraken,* p. 274-5; Mackal, *Searching,* p. 38.

8. Ellis, *Monsters,* pp. 303-4.

9. Ibid., p. 304.

10. Addison Verrill, "A gigantic Cephalopod on the Florida coast. *American Journal of Science* ser. 4, 3 (January 1897): 79.

11. Heuvelmans, *Kraken,* p. 274.

12. Ellis, *Monsters,* p. 306.

13. Ibid.

14. Ibid.

15. Addison Verrill, "Additional information concerning the giant Cephalopod of Florida." *American Journal of Science* ser. 4, 3 (February 1897): 162-163.
16. Ellis, *Monsters,* p. 307.

17. Mackal, *Searching,* p. 42; Addison Verrill, "The Florida Sea-Monster," *American*

Naturalist 31 (April 1897): 304-7.

18. Ellis, *Monsters,* p. 308.
19. Ibid., pp. 308-9.

20. Addison Verrill, "The supposed great Octopus of Florida; certainly not a Cephalopod. *American Journal of Science* ser. 4, 3 (April 1897): 355-356.

21. Ibid.

22. Ellis, *Monsters,* p. 310; Heuvelmans, *Kraken,* pp. 275-6.

23. "St. Augustine Monster," Wikipedia, http://en.wikipedia.org/wiki/St._Augustine_Monster; Mackal, *Searching,* pp. 34-5.

24. Mackal, *Searching,* pp. 35-6, 42-3.

25. Wood and Gennaro.

26. Mackal, "Biochemical Analyses," pp. 56-60.

27. Ibid., pp. 60-1.

28. Mackal, *Searching,* pp. 39-41; "St. Augustine Monster."

29. Ellis, *Monsters,* pp. 319-20.

30. Pierce et al. (1995), pp. 220-5.

31. Ibid., pp. 228-9.

32. Ellis, *Monsters,* p. 322.

33. Pierce et al. (2004).

34. Ibid.

35. Ibid.

36. Broad.

Chapter 4

1. Van Der Sluus.

2. Muirhead, p. 24.

3. Young.

4. Costello, pp. 188, 336; "Mjøsa," Wikipedia, http://en.wikipedia.org/wiki/Mj%c3%b8sa; Erik Knatterud, "They tried to kill the sea serpent," Database of Norwegian Sea Serpents, http://www.mjoesormen.no/theytriedtokilltheseaserpent.htm.

5. Guiley, p. 137; "Narwhal," Wikipedia, http://en.wikipedia.org/wiki/Narwhal.

6. Holland, "Beached leviathans"; Holland, "Herring Hoggs and Sea Devils."

7. Harold Wilkins, *Secret Cities of Old South America* (New York: Library Publishers, 1952), pp. 328-9.

8. Ibid.; "Church of Santo Domingo de Guzmán," Wikipedia, http://en.wikipedia.org/wiki/Santo_Domingo,_Oaxaca.

9. Heuvelmans, *Wake,* p. 587; Roesch, "Review" (Autumn 1997), p. 7.

10. A.C. Oudemans, *The Great Sea-Serpent* (Leiden: Brill, 1892), p. 58.

11. Heuvelmans, *Wake,* p. 587; Directory of Cities and Towns in Rogaland Fylke, Norway, http://www.fallingrain.com/world/NO/14/a/K; "List of towns and cities in Norway," Wikipedia, http://en.wikipedia.org/wiki/List_of_towns_and_cities_in_Norway; "Karmøy," Wikipedia, http://en.wikipedia.org/wiki/Karm%C3%B8y.

12. Oudemans, p. 58; Heuvelmans, *Wake,* p. 587; Roesch, "Review" (Autumn 1997), pp. 7-8.

13. Ellis, *Search,* p. 18; "Sogn og Fjordane," Wikipedia, http://en.wikipedia.org/wiki/Sogn_og_Fjordane; "List of villages in Sogn og Fjordane," Wikipedia, http://en.wikipedia.org/wiki/List_of_villages_in_Sogn_og_Fjordane.

14. Ellis, *Search,* p. 18; Heuvelmans, *Wake,* p. 587; Roesch, "Review" (Autumn 1997), pp. 7-8.

15. Ellis, *Search,* p. 18; Heuvelmans, *Wake,* p. 587; Roesch, "Review" (Autumn 1997), pp. 7-8.

16. Holland, "Beached leviathans."

17. Ibid.

18. "Basking Shark," U.S. Department of the Interior, http://www.gma.org/fogm/cetorhinus_maximus.htm.

19. Heuvelmans, *Wake,* p. 118; Shuker, "Bring Me the Head."

20. Heuvelmans, *Wake,* p. 118; Shuker, "Bring Me the Head."
21. Heuvelmans, *Wake,* p. 119.

22. Ibid.

23. Ibid., pp. 119-20.

24. Ibid., pp. 120-3.

25. Ibid., pp. 123-5; Dinsdale, pp. 184-6; Roesch, "Review" (Autumn 1997), pp. 8-12; Shuker, "Bring Me the Head."

26. Smith, pp. 1-2.

27. Ibid., p. 10.

28. Ibid., pp. 10-12.

29. Costello, pp. 190-1.

30. Heuvelmans, *Wake,* p. 248.

31. Ibid., pp. 248-9.

32. Ibid., p. 250.

33. Ibid.; Roesch, "Review" (Autumn 1997), pp. 19-20.

34. Heuvelmans, *Wake,* p. 587; Roesch, "Review" (Autumn 1997), p. 21.

35. "Loch Ness Monster," Cryptopedia, http://www.chateaugrrr.com/cryptopedia/loch-ness-monster; Gary Campbell, "Ness: The Same Old Story," Big Cats in Britain, http://www.bigcatsinbritain.org/sameoldnesstory.htm.

36. "Capt. Ingall's Story: The Dead Sea-Serpent he saw off Monhegan Island," *New York Times,* 10 June 1880.

37. Heuvelmans, *Wake,* p. 580; Roesch, "Review" (Autumn 1997), p. 22; "Humpback whale," Wikipedia, http://en.wikipedia.org/wiki/Humpback_Whale.

38. "An Unnamed Sea Monster," *New York Times,* 14 December 1881.

39. "The Marlborough Sea-Serpent," *New York Times,* 21 December 1881.

40. E.D. Cope, "The Fossil Reptiles of New Jersey," *The American Naturalist*, 1 (1868): 23–30.

41. Heuvelmans, *Wake,* p. 587; Roesch, "Review" (Winter-Spring 1998), p. 25; "Otago Region," Wikipedia, http://en.wikipedia.org/wiki/Otago; "Oarfish," Wikipedia, http://en.wikipedia.org/wiki/Oarfish.

42. Fort, p. 622.

43. Heuvelmans, *Wake,* p. 587; Roesch, "Review" (Winter-Spring 1998), p. 26.

44. Roesch, "Review" (Winter-Spring 1998), p. 26.

45. Fort, p. 609.

46. Heuvelmans, *Wake,* pp. 416-17.

47. Ibid., p. 417.

48. Ibid., p. 425; Shuker, *In Search of Prehistoric Survivors,* pp. 126-7; Roesch, "Review" (Winter-Spring 1998), p. 31.

49. Heuvelmans, *Wake,* p. 131.

50. "1885 Atlantic hurricane season," Wikipedia, http://en.wikipedia.org/wiki/1885_Atlantic_hurricane_season.

51. "New River (North Carolina)," Wikipedia, http://en.wikipedia.org/wiki/New_River_(North_Carolina); "New River (Broward County, Florida)," Wikipedia, http://en.wikipedia.org/wiki/New_River_(Broward_County,_Florida); Roesch, "Review" (Winter-Spring 1998), p. 33; "Fort Lauderdale, Florida," Wikipedia, http://en.wikipedia.org/wiki/Fort_Lauderdale,_Florida.

52. Heuvelmans, *Wake,* p. 131; Roesch, "Review" (Winter-Spring 1998), p. 33; "Basking Shark," Florida Museum of Natural History, http://www.flmnh.ufl.edu/fish/Gallery/Descript/baskingshark/baskingshark.html.

53. Heuvelmans, *Wake,* pp. 587, 624; Roesch, "Review" (Winter-Spring 1998), p. 33.

54. Heuvelmans, *Wake,* p. 587.

55. Ibid.

56. Mike Dash, "Of Giant Eeels," Charles Fort Institute, http://blogs.forteana.org/node/114; Dash, "Lake Monsters: Status Report."

57. Heuvelmans, *Wake,* p. 112.

58. Ibid., pp. 112-13.

59. Roesch, "Review" (Winter-Spring 1998), p. 34.

60. Ibid.

61. Heuvelmans, *Wake,* pp. 587, 624.

62. Ibid., p. 413.

63. Ibid.

64. Ibid., p. 365; Roesch, "Review" (Summer 1998), p. 27.

65. Heuvelmans, *Wake,* p. 365; "Baird's Beaked Whale," American Cetacean Society, http://www.acsonline.org/factpack/BairdsWhale htm.

66. "Arnoux's beaked whale," African Marine Animals, http://csiwhalesalive.org/csiarnouxs.html; *"Berardius arnuxii," Animal Diversity Web, http://animaldiversity.ummz.umich.edu/site/accounts/information/Berardius_arnuxii.html.*

67. Dash, "Of Giant Eeels"; Dash, "Lake Monsters: Status Report."

68. F.W. Holiday, *The Great Orm of Loch Ness* (London: Faber, 1971), p.172.

69. Heuvelmans, *Wake,* p. 85.

70. Brianna Bailey, "A Look Back: 'Monster' seen in Newport, but was it one?" Newport Beach (CA) *Daily Pilot,* 13 March 2010.

71. Heuvelmans, *Wake,* p. 588.

72. Ibid., pp. 588, 624.

73. Roesch, "Review" (Summer 1998), pp. 30-1.

74. Heuvelmans, *Wake,* p. 588.

75. "The oarfish: 'sea serpent' remains mystery of science," Woods Hole Oceanographic Institution,http://www.thefreelibrary.com/The+oarfish%3A+%22sea+serpent% 22+remains+mystery+of+science.-a013977168.

76. Fort, p. 622.

77. "A sea monster," London *Daily News*, 26 June 1908; "William Plane Pycraft," Wikipedia, http://en.wikipedia.org/wiki/William_Plane_Pycraft.

78. Heuvelmans, *Wake*, pp. 588, 624; Roesch, "Review" (Winter-Spring 1999), pp. 15-16.

79. "Photo in the News: Dead Whale Found With Car-Size Tongue," National Geographic Nes (18 July 2007), http://news.nationalgeographic.com/news/2007/07/070718-whale-tongue.html; "Blue whale," Wikipedia, http://en.wikipedia.org/wiki/Blue_whale.

80. "Mysterious lake creature shrouded in myth"; "Ogopogo"; Chorvinsky, "Nessie"; "See Creatures?"

81. "Mysterious lake creature shrouded in myth."

82. Costello, p. 223.

83. Guiley, p. 133.

84. "Champlain 'Sea Serpent," *New York Times,* 19 April 1915.

85. "Find Huge Sea Monster," *New York Times,* 13 February 1921.

86. Roesch, "Review" (Winter-Spring 1999), pp. 17-18; Coleman and Huyghe, p. 281.

87. Heuvelmans, *Wake,* p. 406.

88. Ibid.; Roesch, "Review" (Winter-Spring 1999), p. 19; Systematic Marine Biodiversity System, http://symbiosis.nre.gov.my/Species/Pages/Globicephala%20macrorhynchus.aspx.

89. F.A. Mitchell-Hedges, *Battles with Great Fish* (London: Duckworth, 1923), p. 22.

90. "Elephant seal," Wikipedia, http://en.wikipedia.org/wiki/Elephant_seal.

91. Fort, pp. 621-2.

92. Heuvelmans, *Wake,* p. 588; Roesch, "Review" (Winter-Spring 1999), pp. 19-20.

93. Roesch, "Review" (Winter-Spring 1999), p. 20.

94. Ibid., pp. 20-1.

95. Shuker, "Trunko - Two More Photographs!!"

96. Fort, p. 622.

97. Roesch, "Review" (Winter-Spring 1999), p. 21; Heuvelmans, *Kraken,* p. 238; Heuvelmans, *Wake,* p. 73.

98. Heuvelmans, *Kraken,* pp. 238-9.

99. Heuvelmans, *Wake,* pp. 73, 569, 588, 624.

100. "The Legend of Trunco [*sic*]." Margate Business Association, http://www.margatebusiness.co.za/index.php?option=com_content&view=article&id=64:the-legend-of-trunco&catid=1:mba-news&Itemid=2.

101. Ibid.

102. Roesch, "Review" (Winter-Spring 1999), p. 21; Heuvelmans, *Wake,* p. 588; Lance Bradshaw, "Trunko," Kryptid's Keep, http://www.angelfire.com/sc2/Trunko/trunko.html.

103. Heuvelmans, *Wake,* p. 465; Chorvinsky, "Gallery of Globsters."

104. Chorvinsky, "Gallery of Globsters."

105. Ibid.

106. Heuvelmans, *Wake,* p. 465.

107. Ibid., pp. 135, 588.

108. Harrison, *Sea Serpents,* pp. 104-5.

109. Harrison, *Sea Serpents,* pp. 105-6; Heuvelmans, *Wake,* pp. 135, 588.

110. "Reptile's Fossil Found," *New York Herald Tribune,* 16 June 1928.

111. Heuvelmans, *Wake,* p. 588; "Narwhal," Wikipedia, http://en.wikipedia.org/wiki/Narwhal.

112. Heuvelmans, *Wake,* p. 484; Shuker, "Bring Me the Head."

113. Heuvelmans, *Wake,* p. 484.

114. Shuker, "Bring Me the Head"; "Notacanthidae," Wikipedia, http://en.wikipedia.org/wiki/Notacanthidae.

115. "Ice Bares Strange Animal," *New York Times,* 26 November 1930.

116. Fort, pp. 623-4; Chorvinsky, "Gallery of Globsters."

117. "Scientific Riddle; This Furry Monster of a Million Years Ago, Brought to Shore by His Glacial Tomb," *Ogden Standard-Examiner*, 11 January 1931.

118. Ibid.; Heuvelmans, *Wake*, pp. 570, 588; Chorvinsky, "Gallery of Globsters."

119. Shuker, "Son of Trunko!"

120. Heuvelmans, *Wake*, pp. 570, 588.

121. Ibid., p. 132.

122. Ibid., pp. 133-4.

123. Smith, p. 57.

124. Ibid.

125. "Henry Island (Washington)," Wikipedia, http://en.wikipedia.org/wiki/Henry_Island_(Washington); "San Juan County, Washington," Wikipedia, http://en.wikipedia.org/wiki/San_Juan_County,_Washington.

126. "Monster is found dead," *Kingsport* (Tenn.) *Times*, 22 November 1934.

127. "Monster found by fisherman mystifies Canadian scientists," *Gettysburg* (Pa.) *Times*, 23 November 1934; "Huge sea serpent is 'Zaweaksh' of olden times, Indian sages maintain," *Port Arthur* (Texas) *News*, 25 November 1934.

128. Heuvelmans, *Wake*, pp. 131-2; LeBlond and Bousfield, pp. 46-8.

129. "Unknown animal washed ashore on beach," *Sidney Morning Herald*, 16 April 1935.

130. Smith, pp. 47-8.

131. "Twofold Bay," Wikipedia, http://en.wikipedia.org/wiki/Twofold_Bay.

132. "Strange marine animal found on beach," *Sidney Morning Herald*, 4 May 1935.

133. Smith, p. 48.

134. Karl Shuker, *The Unexplained* (London: Carlton, 1996), p. 224.

135. LeBlond and Bousfield, pp. 14-25.

136. Naish, "Another Caddy Carcass?"

137. LeBlond and Bousfield, pp. 50-1.

138. Ibid., pp. 52-3.

139. Ibid., pp. 47, 55-7.

140. Heuvelmans, *Wake,* pp. 288, 624.

141. Henry Bigelow and William Schroeder, "Sharks," in J. Tee-Van, C.M. Breder, S.F. Hildebrand, A.E. Parr and W.C. Schroeder (eds.) *Fishes of the Western North Atlantic: Part One* (New Haven, CT: Yale University Press, 1948), p. 147.

142. LeBlond and Bousfield, p. 48.

143. Ibid., pp. 48-9.

144. Douglas Sutherland, *Against the Wind an Orkney Idyll* (London: Cox & Wyman, 1966), p. 164.

145. J.G. Marwick, "Nature Notes," *The Orcadian* (Kirkwall, Orkney), 29 January 1942.

146. Ibid.

147. J.G. Marwick, "Nature Notes," *The Orcadian,* 5 February 1942.

148. "More about the monster - declared a shark by south," *The Orcadian,* 5 February 1942.

149. Dinsdale, pp. 174-6.

150. Ibid., p. 177.

151. Harrison, *Sea Serpents,* pp. 127-8.

152. Ibid., pp. 128-9.

153. Shuker, "Bring Me the Head."

154. "Scots see sea monster," *New York Times,* 3 October 1944; "Furry Beast of Machrihanish 1944," The Kintyre Forum, http://www.kintyreforum.com/viewtopic.php?f=22&t=10837&start=0&sid=894620bdcf1f9d20854a9c1213d04e15; Heuvelmans, *Wake,* pp. 570, 588.

155. "Prehistoric Monster Found in Alaska," *Traverse City Record-Eagle,* 25 October 1946.

156. Heuvelmans, *Wake,* pp. 588, 624.

157. "Risso's dolphin," Wikipedia, http://en.wikipedia.org/wiki/Risso%27s_Dolphin; "Killer Whales," Sea World, http://www.seaworld.org/animal-info/info-books/killer-whale/physical-characteristics.htm.

158. Heuvelmans, *Wake,* pp. 474-5, 588, 621-2.

159. Ibid., p. 475.

160. Ibid.; LeBlond and Bousfield, pp. 49, 120.

161. "Dunk Island," Wikipedia, http://en.wikipedia.org/wiki/Dunk_Island.

162. "Globster," Cryptopedia, http://www.chateaugrrr.com/development/cryptopedia/globster.

163. Heuvelmans, *Wake,* p. 588; "Ataka Carcass," American Monsters, http://americanmonsters.com/site/2010/01/ataka-carcass-egypt.

164. J.D. Adams, "Oregon Mysteries of the Sea," http://www.travel-to-oregon-tips.com/mystery-sea.html; "Sea Serpents and Lake Monsters: Legends and Myths, or Reality?" The Shadowlands, http://theshadowlands.net/serpent.htm.

164. "Sea Serpents and Lake Monsters: Legends and Myths, or Reality?"

166. Peter Hassall, "50 years ago this month," *Fortean Times* 142 (February 2001): 18.

167. Heuvelmans, *Wake,* pp. 141-2.

168. Ibid., p. 588.

167. Smith, p. 48.

170. Heuvelmans, *Wake,* p. 588.

171. Peter Costello, "Canvey Island monsters," *Fortean Times* 242 (December 2008): 71; Gary Hammond, "Canvey Island monsters," *Fortean Times* 245 (March 2009): 73.

172. Dinsdale, pp. 180-2; Heuvelmans, *Wake,* p. 135.

173. Heuvelmans, *Wake,* p. 588.

174. Costello, "Canvey Island monsters"; Hammond, "Canvey Island monsters."
175. Dunning, "Attack of the Globsters!"; "Globster," Wikipedia.

176. The Alaska Monster List, http://s8int.com/WordPress/?tag=cryptozoology.

177. "Mystery of a Monster," *Life* (6 August 1956): 38; The Alaska Monster List.

178. Heuvelmans, *Wake,* p. 588; Carwardine, pp. 106-7.

179. Picasso, "South American Monsters & Mystery Animals," p. 29.

180. Heuvelmans, *Wake,* pp. 140-1.

181. "Treasure Island Florida History," http://treasureislandflorida.org/history.htm.

182. Heuvelmans, *Wake,* p. 588.

183. Sanderson, *"Things,"* pp. 11-12; Smith, pp. 55-6.

184. Sanderson, *"Things,"* p. 12; Smith, p. 56.

185. Sanderson, *"Things,"* pp. 13-14; Smith, p. 56.

186. "Monster to get thorough check," *Hobart Mercury,* 19 March 1962.

187. Sanderson, *"Things,"* p. 13; Ellis, *Monsters of the Sea,* pp. 315-16; Smith, p. 56.

188. Heuvelmans, *Wake,* p. 588.

189. Dinsdale, p. 183.

190. Heuvelmans, *Wake,* p. 588.

191. LeBlond and Bousfield, p. 120.

192. Stanton Wood, "Coney Island Globster," The Unbelievably Strange Wildlife Garden, http://treesquid.blogspot.com/search/label/Coney%20Island%20Globster; John Devlin, "Aquarium shows giant octopus armed with 1,600 suction cups," *New York Times,* 28 March 1962.

193. Wood, "Coney Island Globster."

194. Heuvelmans, *Wake,* p. 588.

195. Ibid.; Cheryl Lynn Dybas, "The oarfish: 'sea serpent' remains mystery of science," CBS Interactive Business Network, http://findarticles.com/p/articles/mi_hb3324/is_n1_v36/ai_n28625033.

196. LeBlond and Bousfield, pp. 49-50.

197. "Weird creature on Orleans beach looks like sea-serpent," *Cape Codder,* 24 December 1964.

198. Clark, *Unexplained!,* p. 2511; Eberhart, p. 209.

199. Meurger and Gagnon, p. 110.

200. Adams, "Oregon Mysteries of the Sea"; James Adams personal correspondence; Peter Cairns, "Colossal Claude and the sea monsters," *The Oregonian,* 24 September 1967.

201. Chorvinsky, "Gallery of Globsters."

202. Ibid.

203. "The Mann Hill Monster," Cryptozoo, http://dagmar.lunarpages.com/~parasc2/en/cryptozoo/aquarium08.htm.

204. Smith, pp. 59-60.

205. Harrison, *The Encyclopedia of the Loch Ness Monster,* pp. 63-4.

206. Downes; Harrison, *Sea Serpents,* p. 96.

207. Downes.

208. Bowden; Shuker, "Bring Me the Head."

209. Sasaki et al., pp. 45-83.

210. Ibid., p. 48.

211. Ibid., p. 63-66.

212. Bowden.

213. Sasaki et al., p. 65.

214. Bowden; John Koster, "What was the New Zealand monster?" Missouri Association for Creation, http://www.gennet.org/facts/nessie.html.

215. Bowden.
216. "The Best Beaches in Gambia," Beach Holiday Guide, http://www.beachholidayguide.co.uk/gambia_best_beaches.php; "Bungalow Beach Hotel," Trip Advisor,http://www.tripadvisor.com/Hotel_Review-g480198-d316654-Reviews-Bungalow_Beach_Hotel-Serekunda.html; Coleman and Huyghe, p. 133.

217. Shuker, "Bring me the head"; Coleman and Huyghe, p. 134.

218. Coleman and Huyghe, p. 134.

219. Ibid., p. 133; . Shuker, "Bring me the head."

220. Naish, "A Russian sea monster."

221. Ibid.

222. Ibid.

223. Ellis, *Monsters of the Sea,* pp. 317, 363.

224. Pierce et al. (1995), p. 229; Coleman and Huyghe, p. 285; Pierce et al. (2004), p. 125.

225. "Globster," Wikipedia; Dunning, "Attack of the Globsters"; "St. Augustine Monster and Related Globsters," Crypto Web,

http://www.fortunecity.com/roswell/siren/552/marine_globster.html.

226. "The beast of Benbecula"; Clark, *Unexplained!,* p. 251.

227. "The beast of Benbecula."

228. Ibid.

29. Michael Cenedlla, "20th & 21st Century Dinosaurs - Sea Serpent? Plesiosaur?" http://s8int.com/dino28.html.

230. "Bermuda Blob 2, Wikipedia, http://en wikipedia.org/wiki/Bermuda_Blob_2; ; "St. Augustine Monster," Wikipedia, http://en.wikipedia.org/wiki/St._Augustine_Monster#1995_analysis.

231. Ellen Jane Hollis, personal communication with the author, 17 February 2011; Pierce et al (2004), p. 126.

232. "Is a Giant Eel Stalking Loch Ness?" PR Web, http://www.prweb.com/releases/2005/06/prweb247668.htm.

233. "The Loch Ness Tooth," The Museum of Hoaxes, http://www.museumofhoaxes.com/hoax/archive/permalink/the_loch_ness_tooth; Dash, "Lake Monsters: Status Report," p. 30.

234. "Dragon, ahoy!"; "From the Vault," *Fortean Times* 209 (June 2006): 80.

235. "Dragon ahoy!"; Coleman and Huyghe, p. 284.

236. "Good month for monster hunters"; Dubious Globsters, http://www.geocities.com/ capedrevenger/dubiousglobsters.html; Coleman and Huyghe, pp. 284-5.

237. "Good month for monster hunters"; Coleman and Huyghe, p. 285.

238. Coleman and Huyghe, p. 285; "Good month for monster hunters"; "; Dubious Globsters, http://www.geocities.com/capedrevenger/dubiousglobsters.html.

239. "Hunka, hunka, stinking globster."

240. Ibid.

241. Ibid.

242. Pierce et al (2004), p. 127; L. Thurber personal correspondence with the author, ** February 2011.

243. Pierce et al (2004), pp. 127, 130.

244. Hemmler; "Claveria, Masbate," Wikipedia, http://en.wikipedia.org/wiki/ Claveria,_Masbate.

245. Hemmler.

246. Ibid.; Coleman and Huyghe, pp. 279-80.

247. Bermuda *Royal Gazette,* 13 and 14 January 1997 and 14 June 1997; Pierce et al (2004), p. 133.

248. "Al-Fantas [*sic*] Fishman," http://www.fortunecity.com/roswell/siren/552/ marine_fishstory.html.

249. Ibid.

250. Ibid.; Coleman and Huyghe, p. 283.

251. "Mysterious sea creature washes up on New Zealand beach," CNN (14 October 1997), http://www.cnn.com/EARTH/9710/14/new.zealand.creature/index.html.
252. Ben Roesch, "Other Cryptozoology News," *The Cryptozoology Review* 2 (Winter-Spring 1998): 11-12; "Four Mile Blobster," American Monsters, http://americanmonsters.com/ site/2010/01/four-mile-blobster-tasmania;Colemanand Huyghe, pp. 285-6.

253. Shuker, "The great sea serpent visits Greatstone."

254. "Newfoundland Blob," Wikipedia, http://en.wikipedia.org/wiki/Newfoundland_Blob; Carr et. al.

255. "Giant blob baffles marine scientists," BBC News, 2 July 2003; "Chilean blob could be an octopus," BBC News, 3 July 2003.

256. "Chilean blob could be an octopus."

257. Pierce et al. (2004), pp. 126, 130; Broad.

258. "The sea monster of Parton," *Whitehaven News,* 22 February 2006.

259. Ibid.; "There's nowt new about 'sea pigs," *Whitehaven News,* 24 September 2004.

260. "The sea monster of Parton."

261. "Carcass of unidentified animal on the beach Nea Kydonia," http://travel.webshots.com/photo/2657682110060412234hSUILr.

262. "Hellish hairy sea monster cast ashore." Pravda, http://english.pravda.ru/photo/album/sea_monster-1816.

263. "Aliens land in Russia," Pravda, http://english.pravda.ru/news/society/20-01-2010/111738-aliens-0; "Weird hairy females seduce hot-blooded Caucasian men," Pravda, http://english.pravda.ru/society/anomal/17-12-2008/106840-weird_females-0.

264. Joye Brown, "The Montauk Monster: Legend or latex?" *Newsday,* 31 July 2008.

265. "The Hound of Boncaville," *The Independent,* 23 July 2008; "Montauk residents proud of their 'monster,'" *Newsday,* 31 July 2008.

266. "The Hound of Boncaville"; Darren Naish, "What was the Montauk monster?" Tetrapod Zoology, http://scienceblogs.com/tetrapodzoology/2008/08/the_montauk_monster.php; Hamilton Nolanm "Montauk Monster: Vole or Satan?" Gawker, http://gawker.com/#!5032170/montauk-monster-vole-or-satan; Brown.

267. Loren Coleman, "Alaskan mystery carcass photo."

268. DeMarban.
269. Ibid.

270. Coleman, "Clapsadle Carcass"; Montauk-Monster.

271. Coles; Harris; Naish, "England 'does a Montauk.'"

272. "The revolting 'Dr. Who' sea monster that terrified tourists."

273. "Headless 'thing' washes up on Newfoundland beach."

274. "Kitchenuhmaykoosib monster"; "Bizarre 'corpse' reminiscent of Montauk monster"; Radford, "'Monster' washes ashore in Canada."

275. "Tory Island critter," *Fortean Times* 273 (May 2011): 73.

276. "2010 Montauk Monster discovered!" Montauk-Monster, http://www.montauk-monster.com.

277. Deans.

278. Deans; "Common Brushtail Possum," Wikipedia, http://en.wikipedia.org/wiki/Common_Brushtail_Possum.

Chapter 5

1. Costello, p. 336; "Rogaland," Wikipedia, http://en.wikipedia.org/wiki/Rogaland.

2. Costello, p. 336; Eberhart, p. 668.

3. Eberhart, p. 679.

4. Costello, pp. 156-7.

5. Ibid., p. 136.

6. Ibid., p. 336.

7. Heuvelmans, *Wake,* p. 418.

8. Ibid., pp. 418, 575.

9. Karatzas.

10. Ibid.

11. O'Neill, pp. 54-6.

12. Heuvelmans, *Wake,* p. 130.

13. USA Place Names, http://www.placenames.com/us.

14. Guiley, p. 132.

15. Heuvelmans, *Wake,* pp. 85-6.

16. Oudemans, p. 80.

17. Ibid.

18. Ibid., pp. 80, 82.

19. Ibid., p. 79.

20. Ibid.; *"Lineus longissimus,"* Wikipedia, http://en.wikipedia.org/wiki/Lineus_longissimus.

21. Heuvelmans, *Wake,* p. 227; Roesch, "A Review of Alleged Sea Serpent Carcasses Worldwide, Part 1," p. 16; Coleman and Huyghe, p. 284; Shuker, "Bring Me the Head."

22. Heuvelmans, *Wake,* pp. 227-9.

23. Roesch, "A Review of Alleged Sea Serpent Carcasses Worldwide, Part 1," p. 16; Dinsdale, p. 166.

24. Clark, *Unnatural Phenomena,* p. 348.

25. "Common Mudpuppy," Wikipedia, http://en.wikipedia.org/wiki/Common_Mudpuppy.

26. Oudemans, pp. 84-5.

27. Ibid., p. 85; Heuvelmans, *Wake,* pp. 235-6, 587.

28. Heuvelmans, *Wake,* pp. 85, 587.

29. Costello, pp. 215, 336.

30. *New York Times,* 12 May 1877.

31. *New York Times,* 14 May 1877.

32. Oudemans, p. 86.
33. Ibid.

34. Heuvelmans, *Wake,* pp.138-40.

35. Ibid.; Shuker, "Bring Me the Head"; Michael Bright, *There Are Giants in the Sea* (London: Robson, 1989), pp. 185-6; "*Chlamydoselachus*," Wikipedia, http://en.wikipedia.org/wiki/Chlamydoselachus.

36. Roesch, "A Review of Alleged Sea Serpent Carcasses Worldwide, Part 1," pp. 24-5; "Crestfish," Wikipedia, http://en.wikipedia.org/wiki/Crestfish.

37. *New York Times,* 23 September 1880.

38. Roesch, "A Review of Alleged Sea Serpent Carcasses Worldwide, Part 1," p. 26.

39. *New York Times,* 15 September 1883.

40. Roesch, "A Review of Alleged Sea Serpent Carcasses Worldwide, Part 2," pp. 26-7.

41. "A Bolivian Saurian," *Scientific American* 49 (1883): 3.

42. Clark, *Unnatural Phenomena,* pp. 129-30.

43. Dash, "Lake Monsters: Status Report," p. 30.

44. Clark, *Unnatural Phenomena,* p. 45.

45. Ibid., p. 55.

46. "Snakehead (fish)," Wikipedia, http://en.wikipedia.org/wiki/Channidae.

47. Heuvelmans, *Wake,* p. 140.

48. Ibid., pp. 140, 587.

49. USA Place Names.

50. Eberhart, p. 680.

51. "Takes curious fish out of the lake," *Milwaukee Journal,* 10 July 1902.

52. "The sea serpent: a report from the Copelands," *County Down Spectator,* 11 September 1908.

53. Roesch, "A Review of Alleged Sea Serpent Carcasses Worldwide, Part Four," pp. 16-17.
54. "Marvelous fish," *Galveston Daily News,* 21 August 1910.

55. Clark, *Unnatural Phenomena,* p. 30.

56. LeBlond and Bousfield, p. 57.

57. Ibid., pp. 57-9.

58. "Tales from the Vault."

59. LeBlond and Bousfield, pp. 59, 121.

60. "Cryptoid Malaysia."4

61. Morrissey.

62. Ibid.

63. Ibid.

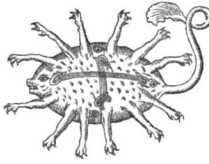

Bibliography

Books

- Carwardine, Mark. *Whales, Dolphins and Porpoises.* London: Dorling Kindserley, 1995.
- Clark, Jerome. *Unexplained: Strange Sightings, Incredible Occurrences, and Puzzling Physical*
- *Phenomena* 2d edition. Canton, MI: Visible Ink Press, 2003.
- - . *Unnatural Phenomena: A Guide to the Bizarre Wonders of North America.* Santa Barbara, CA: ABC-CLIO, 2005
- Coleman, Loren, and Patrick Huyghe. *The Field Guide to Lake Monsters, Sea Serpents, and Other Mystery Denizens of the Deep.* New York: Tarcher/Penguin, 2003.
- Costello, Peter. *In Search of Lake Monsters.* New York: Coward-McCann and Geoghegan, 1974.
- Dinsdale, Tim. *Monster Hunt.* Washington, DC: Acropolis, 1972.
- Eberhart, George. *Mysterious Creatures: A Guide to Cryptozoology.* Santa Barbara, CA: ABC-CLIO, 2002.
- Ellis, Richard. *Men & Whales.* Lyons, NY: Lyons Press, 1999.
- - . *Monsters of the Sea.* New York: Alfred A. Knopf, 1994.
- - . *The Search for the Giant Squid: The Biology and Mythology of the World's Most Elusive Sea Creature.* New York: Penguin, 1990.
- Fort, Charles. *The Complete Books of Charles Fort.* Mineola, NY: Dover Publications, 1975.
- Gaal, Arlene. *Ogopogo: The True Story of the Okanagan Lake Million Dollar Monster.* Surrey, BC: Hancock House, 1986.
- Gould, Rupert. *The Case for the Sea-Serpent.* London: Phillip Allen, 1930.
- Guiley, Rosemary. *Atlas of the Mysterious in North America.* New York: Facts on File, 1995.
- Harrison, Paul. *The Encyclopedia of the Loch Ness Monster.* London: Robert Hale, 1999.
- - . *Sea Serpents and Lake Monsters of the British Isles.* London: Robert Hale, 2001.
- Heuvelmans, Bernard. *In the Wake of Sea Serpents.* New York: Hill and Wang, 1968.
- - . *The Kraken and the Colossal Octopus: In the Wake of Sea-Monsters.* London: Kegan Paul International, 2003.
- *Joint Interim Report, Bahamas Marine Mammal Stranding Event of 15-16 March 2000.* Washington, D.C.: U.S. Department of Commerce, 2001.
- LeBlond, Paul, and Edward Bousfield. *Cadborosaurus: Survivor from the Deep.*

Victoria, BC: Horsdal & Schubart, 1995.

- Mackal, Roy. *Searching for Hidden Animals: An Inquiry into Zoological Mysteries.* London: Cadogan Books, 1980.
- Meurger, Michel, and Claude Gagnon. *Lake Monster Traditions: A Cross-Cultural Analysis.* London: Fortean Tomes, 1988.
- Newton, Michael. *Encyclopedia of Cryptozoology: A Global Guide to Hidden Animals and Their Pursuers.* Jefferson, NC: McFarland, 2005.
- - . *Florida's Unexpected Wildlife.* Gainesville, FL: University Press of Florida, 2007.
- - . *Strange Monsters of the Pacific Northwest.* Atglen, PA: Schiffer, 2011.
- O'Neill, June. *The Great New England Sea Serpent: An Account of Unknown Creatures Sighted by Many Respectable Persons Between 1638 and the Present Day.* Camden, ME: Downeast Books, 1999.
- Oudemans, A.C. *The Great Sea Serpent.* Leiden: E.J. Brill, 1892.
- Sanderson, Ivan. *"Things."* New York: Pyramid Books, 1967.
- Shuker, Karl. *In Search of Prehistoric Survivors.* London: Blandford, 1995.
- Smith, Malcolm. *Bunyips & Bigfoots: In Search of Australia's Mystery Animals.* Alexandria, NSW: Millennium Books, 1996.
- Woodley, Michael. *In the Wake of Bernard Heuvelmans: An Introduction to the History and Future of Sea Serpent Classification.* Bideford, North Devon: CFZ Press, 2008.

Articles

- "The beast of Benbecula." *Fortean Times* 93 (December 1996): 9.
- "Bizarre 'corpse' reminiscent of Montauk monster." *Daily Telegraph* (London): 21 May 2010.
- Broad, William. "Ogre? Octopus? Blobologists Solve an Ancient Mystery." *New York Times* 27 July 2004.
- Carr, S.M., H.D. Marshall, K.A. Johnstone, L.M. Pynn and G.B. Stenson. "How to Tell a Sea Monster: Molecular Discrimination of Large Marine Animals of the North Atlantic."
- *Biology Bulletin* 202 (February 2002): 1-5.
- "Catch of the day." *Fortean Times* 97 (May 1997): 16.
- Chorvinsky, Mark. "Creature on ice." *Strange Magazine* 15 (Spring 1995): 17.
- - . "Monsters on the beach." *Strange Magazine* 15 (Spring 1995): 16.
- Coles, John. "Beast on the Beach." *The Sun* (London): 9 January 2009.
- Costello, Peter. "Canvey Island monsters." *Fortean Times* 242 (December 2008): 71.
- "Cryptoid Malaysia." *Fortean Times* 214 (October 2006): 8-9.
- Deans, Matt. "Creature defies identification." *Coffs Coast Advocate* (New South Wales), 10 November 2010.
- DeMarban, Alex. "Beached carcass may be thing of legends." *Seward* (Alaska) *Phoenix*, 7 August 2008.
- "Devil-fish hunters squids in." *Fortean Times* 187 (June 1996): 19.

- Downes, Jonathan. "Whale of a time with a dragon." *Fortean Times* 95 (February 1997): 46.
- "Dragon, ahoy!" *Fortean Times* 89 (August 1996): 13.
- Dash, Mike. "Lake Monsters: Status Report." *Fortean Times* 102 (September 1997): 28-31.
- "50 years ago this month." *Fortean Times* 142 (February 2001): 18.
- "Five basking sharks found dead on coast," BBC News, 18 June 2004.
- Hammond, Gary. "Canvey Island monsters." *Fortean Times* 245 (March 2009): 73.
- Harris, Paul. "Is this the Beast of Exmoor? Body of mystery animal washes up on beach." *Daily Mail* (London), 9 January 2009.
- Holland, Richard. "Beached leviathans." *Fortean Times* 119 (February 1999): 52.
- Hooker, Sascha, Robin Baird, and Mark Showell. "Cetacean strandings and bycatches in Nova Scotia, eastern Canada, 1991-1996." Paper SC/49/O5 presented to the IWC Scientific Committee, September 1997.
- "Giant squid." *Fortean Times* 149 (September 2001): 17.
- "Giant squid filmed at sea." *Fortean Times* 203 (December 2005): 23.
- Goertzen, John. "New Zuiyo Maru Cryptid Observations." *Creation Research Society Quarterly Journal* 1 (June 2001): 19-29.
- "Good month for monster hunters." *Fortean Times* 95 (February 1997): 18.
- Karatzas, Valerie. "Monstrous fish." *Fortean Times* 184 (July 2004): 76.
- Lucas, Zoe, and Sascha Hooker. "Cetacean strandings on Sable Island, Nova Scotia, 1970-1998." *Canadian Field-Naturalist* 114 (2000): 45-61.
- Mackal, Roy. "Biochemical Analyses of Preserved *Octopus giganteus* Tissue." *Cryptozoology* 5: 55-62.
- MacKenzie, Richard. "Sea monster washes ashore in England." *Mysteries Magazine* 3 (Spring 2005): 8.
- "Moarfish!" *Fortean Times* 196 (June 2005): 14.
- Morrissey, Alisha. "Newfoundland fishermen snag sea monster in nets." *The Telegraph* (St. John's, Newfoundland), 2 March 2010.
- Muirhead, Richard. "Some Chinese Cryptids (Part One)." *Cryptozoology Review* 3 (Winter-Spring 1999): 23-5.
- "Mystery blobby found in Tasmania." *Fortean Times* 109 (April 1998): 21.
- Naish, Darren. "Another Caddy Carcass?" *Cryptozoology Review* 2 (Summer 1997): 26-9.
- - . "Where be Monsters?" *Fortean Times* 132(March 2000): 40-44.
- Norman, S.A., et al. "Cetacean strandings in Oregon and Washington between 1930 and 2002. *Journal of Cetacean Research and Management* 6 (2004): 87-99.
- Picasso, Fabio. "South American Monsters & Mystery Animals." *Strange Magazine* 20 (December 1998): 28-35.
- Pierce, Sidney, Steven Massey, Nicholas Curtis, Gerald Smith Jr., Carlos Olavarria and Timothy Maugel. "Microscopic, Biochemical, and Molecular Characteristics of the Chilean Blob and a Comparison With the Remains of Other Sea Monsters: Nothing but Whales." *Biological Bulletin* 2006 (June 2004): 125-33.

- Pierce, Sidney, Gerald Smith Jr., Timothy Maugel and Eugene Clark. "On the Giant Octopus (*Octopus giganteus*) and the Bermuda Blob: Homage to A.E. Verrill." *Biological Bulletin* 188 (April 1995): 219-30.
- Radford, Benjamin. "'Monster' washes ashore in Canada: Is it the Chupacabra?" *Christian Science Monitor*, 24 May 2010.
- "The revolting 'Dr Who' sea monster that terrified tourists." *Daily Mail* (London), 5 August 2009.
- Roesch, Ben. "Another 'Sea Serpent' Carcass." *Cryptozoology Review* 3 (Winter-Spring 1998):7-8.
- - . "Mystery Carcass off Kuwait." *Cryptozoology Review* 3 (Winter-Spring 1998): 6-7.
- - . "A Review of Alleged Sea Serpent Carcasses Worldwide." *Cryptozoology Review* 2 (Autumn 1997): 6-27; 2 (Winter-Spring 1998): 25-35; 3 (Summer 1998): 27-31; 3 (Winter-Spring 1999): 15-22.
- Sasaki, T., F. Yasuda, K. Nasu, and Y. Taki (eds.). *Collected Papers on the Carcass of an Unidentified Animal trawled off New Zealand by the Zuyo-maru*. Tokyo: La Société Franco–Japonaise d'Océanographie, 1978.
- "Second shark stranding at New York City beach in 2 weeks." *China Post*, 17 September 2007.
- Shuker, Karl. "Bring Me the Head of the Sea Serpent!" *Strange Magazine* 15 (Spring 1995): 12-17.
- - . "The Chilean globster." *Fortean Times* 175 (November 2003): 58-9.
- - . "A colossal surprise." *Fortean Times* 226 (September 2007): 24.
- - . "The enormous hype over the colossal squid." *Fortean Times* 172 (August 2003): 56.
- - . "Globster identity confirmed." *Fortean Times* 189 (December 2004): 20.
- - . "The great sea serpent visits Greatstone." *Fortean Times* 114 (September 1998): 18-19.
- - . "In a flap over a muscular mollusc." *Fortean Times* 163 (November 2002): 21.
- - . "A new home for the Florida globster." *Fortean Times* 175 (November 2003): 59.
- - . "St. Augustine globster research results announced." *Strange Magazine* 15 (Spring 1995):14.
- - . "Squid's in." *Fortean Times* 138 (October 2000): 18.
- Stroud, Richard. "Pathobiology of Marine Mammal Strandings Along the Pacific Coast, 1976-1977." International Association for Aquatic Animal Medicine 1977 Proceedings: 29-30.
- "Tales from the Vault." *Fortean Times* 196 (June 2005): 80; 209 (June 2006): 80.
- Van der Sluus, Marinus. "Korean signs and wonders." *Fortean Times* 223 (July 2007): 56-7.
- Witze, Alexandra. "DNA tests solving sea mysteries." *Dallas Morning News*, 20 March 2002.
- Wood, Forrest, and Joseph Gennaro Jr. "An Octopus Trilogy." *Natural History* 80 (March 1971): 15-24, 84-7.
- Young, Simon. "The Sea Giantess of Ireland." *Fortean Times* 226 (September 2007): 58-9.

Internet Sources

- Adams, J.D. "Oregon Mysteries of the Sea," http://www.travel-to-oregon-tips.com/mystery-sea.html.
- "The Alaska Monster List," http://s8int.com/WordPress/?tag=cryptozoology.
- "Andrew's beaked whale," Wikipedia, http://en.wikipedia.org/wiki/Andrews%27_Beaked_Whale.
- "Basking shark strands in Cornwall," Wildlife Extra, http://www.wildlifeextra.com/go/news/starnded-basking-shark.html#cr.
- "Beached whale," Wikipedia, http://en.wikipedia.org/wiki/Beached_whale.Bowden, Malcolm. "The Japanese Carcass: A Plesiosaur-Type Animal!"
- Creation Start Page,http://www.mbowden.surf3.net/plsfin13.htm.
- "Cetacean Stranding Report," http://projects.earthtech.com/uswtr/EIS/FOEIS-EIS_2009/PDF-FOEIS-EIS/FOEIS-EIS_vol-2/USWTR-FEIS_App-E_Cetacean-Stranding-Report.pdf.
- Cetacean Strandings in the Mediterranean Sea, http://www.cep.unep.org/pubs/meetingreports/MMAP/Cetacean%20Strandings-Ref.pdf.
- "Cetacean Strandings Reported to the California Marine Mammal Stranding Network," National Marine Fisheries Service, http://swr.nmfs.noaa.gov/psd/strand/ceta/2001.htm.
- Chorvinsky, Mark. "Gallery of Globsters." Strange Magazine, http://www.strangemag.com/seaserpgallery.html.
- - . "Nessie and Other Lake Monsters," Strange Magazine, http://www.strangemag.com/ogopogo.html.
- Coleman, Loren. "Alaskan mystery carcass photo," Cryptomundo, http://www.cryptomundo.com/breaking-news/pic-ak-body.
- - . "Clapsdale Carcass: Another Mysterious Bloated Beach Body." Cryptomundo, http://www.cryptomundo.com/cryptozoo-news/another-bod.
- - . "Exmoor Beach Beast," Cryptomundo, http://www.cryptomundo.com/cryptozoo-news/exmoorbeachbeast.
- - . "Globsters' Glory Days Are Gone," Cryptomundo, http://www.cryptomundo.com/cryptozoo-news/globsters-gone.
- "Colossal Squid," Wikipedia, http://en.wikipedia.org/wiki/Colossal_Squid.
- "Cuvier's beaked whale," Wikipedia, http://en.wikipedia.org/wiki Cuvier's_Beaked_Whale.
- "Dana Octopus Squid," Wikipedia, http://en.wikipedia.org/wiki/Taningia_danae.
- "Diamond Squid," Wikipedia, http://en.wikipedia.org/wiki/Thysanoteuthis_rhombus.
- Dolma, Sarah, et al. "A preliminary note on the unprecedented strandings of 45 deep-diving odontocetes along the UK and Irish coast between January and April 2008." http://www.iwcoffice.org/_documents/sci_com/SC60docs/SC-60-E5.pdf.
- Dubious Globsters, http://www.geocities.com/capedrevenger/dubiousglobsters.html.
- Dunning, Bryan. "Attack of the Globsters!" Skeptoid, http://skeptoid.com/episodes/4152.

- Ewing, Bob. "Dolphin, Whale, Porpoise Deaths Increase Study Shows." Digital Journal (7 July 2008), http://www.digitaljournal.com/article/257075
- Dolphin_Whale_Porpoise_Deaths_Increase_Study_Shows.
- "Galiteuthis phyllura," Wikipedia, http://en.wikipedia.org/wiki/Galiteuthis_phyllura.
- "Gervais' beaked whale," Wikipedia, http://en.wikipedia.org/wiki/Gervais%27 27_Beaked_Whale.
- "Giant beaked whale," Wikipedia, http://en.wikipedia.org/wiki/Arnoux% 27s_Beaked_Whale.
- "Giant squid," Wikipedia, http://en.wikipedia.org/wiki/Giant_squid.
- "Ginkgo-toothed beaked whale," Wikipedia, http://en.wikipedia.org/wiki/Ginkgo-toothed_Beaked_Whale.
- "Globster," Monstropedia, http://www.monstropedia.org/index.php?title=Globster.
- "Globster," Wikipedia, http://en.wikipedia.org/wiki/Globster.
- "Globsters!" Strangemag.com, http://www.strangemag.com/globhome.html.
- "Gray's beaked whale," Wikipedia, http://en.wikipedia.org/wiki/Gray% 27s_Beaked_Whale.
- "Hector's beaked whale," Wikipedia, http://en.wikipedia.org/wiki/Hector% 27s_Beaked_Whale.
- Hemmler, Markus. "The Masbate Monster." Still on the Track, http://forteanzoology.blogspot.com/2010/11/markus-hemmler-masbate-monster.html. Hogan, Ron. "Beached Cow Mistaken For Polar Bear,"
- http://www.popfi.com/2010/09/21/beached-cow-mistaken-for-polar-bear.
- Holland, Richard. "Herring Hogs and Sea Devils." *Still on the Track* (17 June 2009), http://forteanzoology.blogspot.com/2009_06_17_archive.html.
- "Humboldt Squid," Wikipedia, http://en.wikipedia.org/wiki/Dosidicus_gigas.
- "Hunka, hunka, stinking globster," Freak-o-pedia, http://www.haxan.com/portfolio/ freakylinks/WWWFRE~1.COM/FREAKO~1/TAILS_~1/GLOBST~1.HTM. "Kitchenuhmaykoosib monster," Wikipedia,http://en.wikipedia.org/wiki/ Kitchenuhmaykoosib_monster.
- "Kondakovia longimana," Wikipedia, http://en.wikipedia.org/wiki/ Kondakovia_longimana.
- "Kraken," Wikipedia, http://en.wikipedia.org/wiki/Kraken.
- Kuchinsky, Charlotte. "Globsters: Dead Sea Monsters or Something Else Altogether?"
- Associated Content, http://www.associatedcontent.com/article/704497/ globsters_dead_sea_monsters_or_something.html.
- Leavitt, Kirk. "A MonsterQuest Look at 'Giant Squid Found,'" http:// www.associatedcontent.com/article/1937344/ a_monsterquest_look_at_giant_squid.html?cat=37.
- "List of Colossal Squid specimens and sightings," Wikipedia, http://en.wikipedia.org/ wiki/List_of_Colossal_Squid_specimens_and_sightings. "List of giant squid specimens and sightings," Wikipedia, http://en.wikipedia.org/wiki/ List_of_giant_squid_specimens_and_sightings.
- "Marine mammals and sonar," Wikipedia, http://en.wikipedia.org/wiki/ Marine_mammals_and_sonar.

- "Malaysian Dragon," http://dagmar.lunarpages.com/~parasc2/en/cryptozoo/aquarium.htm.
- "The Mann Hill Monster," http://dagmar.lunarpages.com/~parasc2/en/cryptozoo/aquarium08.htm.
- "Megalocranchia fisheri," Wikipedia, http://en.wikipedia.org/wiki/Megalocranchia_fisheri.
  "Montauk Monster," Wikipedia, http://en.wikipedia.org/wiki/Montauk_Monster.
- Montauk-Monster: The Truth, The Legend, The Mystery, http://www.montauk-monster.com.
- "Mysterious lake creature shrouded in myth," http://www.hotspotsz.com/Mysterious_lake_creature_shrouded_in_myth_(Article-15406).html.
- Naish, Darren. "England 'does a Montauk.'" Tetrapod Zoology, http://scienceblogs.com/tetrapodzoology/2009/01/england_does_a_montauk.php.
- - . "A Russian sea monster carcass is claimed to be that of an ancient 'archaeocete' whale."
- Tetrapod Zoology, http://scienceblogs.com/tetrapodzoology/2009/03/a_russian_sea_monster_carcass.php.
- "Noise." The Marine Detective, http://www.earthlingenterprises.ca/earthlingenterprises/Noise.html.
- "North Pacific Giant Octopus," Wikipedia, http://en.wikipedia.org/wiki/North_Pacific_Giant_Octopus.
- "Oceania Strandings," Watery World of Whales, http://www.whales.org.au/strandings/oceania/index.html.
- "Ogopogo," Unknown Explorers, http://www.unknownexplorers.com/ogopogo.php.
- Pardo, Priscilla. "Cetacean Strandings from the Caribbean and Pacific Coasts of Costa Rica," http://fundacionpromar.org/docs/Abstracts%20de%20Conferencias.doc.
- Podesta, Michela, and Bruno Cozzi. "Analysis of cetacean strandings along the Italian coastline in the years 1986-2004," http://www.iwcoffice.org/_documents/sci_com/SC58docs/SC-58-O3.pdf.
- "Pygmy beaked whale," Wikipedia, http://en.wikipedia.org/wiki/Pygmy_Beaked_Whale.
- "Robust Clubhook Squid," Wikipedia, http://en.wikipedia.org/wiki/Moroteuthis_robusta.
- Russo, Julie. "In the Octopus's Garden," http://mondopulpo.blogspot.com/2006_05_01_archive.html.
- "Salmon Shark Strandings," http://homepage.mac.com/mollet/Ld/Ld_strandings.html.
- "See Creatures?" The Unexplained, http://www.angelfire.com/planet/thecoldspot/unexplained/creatures.html.
- "Seven-arm Octopus," Wikipedia, http://en.wikipedia.org/wiki/Haliphron_atlanticus.
- "Shepherd's beaked whale," Wikipedia, http://en.wikipedia.org/wiki/Shepherd%27s_Beaked_Whale.
- Shuker, Karl. "Behold, Trunko!!" ShukerNature, http://karlshuker.blogspot.com/2010/09/behold-trunko.html.
- - . "Son of Trunko!" ShukerNature, http://karlshuker.blogspot.com/2010/09/son-of-

trunko.html.

-   - . "Trunko - Two More Photographs!!' ShukerNature, http://karlshuker.blogspot.com/2010/09/trunko-two-more-photographs.html.
- Smith, Dawn. "Whale and Dolphin Mass Strandings," http://www.suite101.com/content/whale-and-dolphin-mass-strandings-a56855.
- "Sowerby's beaked whale," Wikipedia, http://en.wikipedia.org/wiki/Sowerby's_Beaked_Whale.
- "Spade-toothed whale," Wikipedia, http://en.wikipedia.org/wiki/Spade_Toothed_Whale.
- "Stejneger's beaked whale," Wikipedia, http://en.wikipedia.org/wiki/Stejneger%27s_Beaked_Whale.
- Sternfield, Mary. "Marine Mammal Strandings as Reported to National Marine Fisheries Service, Alaska Region: 2003,"http://www.fakr.noaa.gov/protectedresources/strandingrpts/03strandings.pdf.
- "Strange creature found washed up on beach," http://www.hotspotsz.com/Strange_creature_found_washed_up_on_beach_(Article-18542).html.
- "Strap-toothed whale," Wikipedia, http://en.wikipedia.org/wiki/Strap-toothed_Whale. Trivedi, Bijal. "'Weird" New Squid Species Discovered in Deep Sea." National Geographic Today, http://news.nationalgeographic.com/news/2001/12/1220_TVweirdsquid.html.
- "Tropical bottlenose whale," Wikipedia, http://en.wikipedia.org/wiki/Longman%27s_Beaked_Whale.
- "True's beaked whale," Wikipedia, http://en.wikipedia.org/wiki/True%27s_Beaked_Whale.
- "2006, a year of unusual marine strandings." Wildlife Extra, http://www.wildlifeextra.com/go/marine/whales-2006strandings.html#cr.
- West African Cetaceans, http://westafricacetaceans.blogspot.com.
- "Why do sharks strand?" Biology of Sharks and Rays, http://www.elasmo-research.org/education/topics/b_strandings.htm.
- Wood, Stanton. "Coney Island Globster," The Unbelievably Strange Wildlife Garden, http://treesquid.blogspot.com/search/label/Coney%20Island%20Globster.
- "Zuiyo-maru Plesiosaur," http://dagmar.lunarpages.com/~parasc2/en/cryptozoo/aquarium.htm.

STILL ON THE TRACK OF UNKNOWN ANIMALS

The Centre for Fortean Zoology, or CFZ, is a non profit-making organisation founded in 1992 with the aim of being a clearing house for information, and coordinating research into mystery animals around the world.

We also study out of place animals, rare and aberrant animal behaviour, and Zooform Phenomena; little-understood "things" that appear to be animals, but which are in fact nothing of the sort, and not even alive (at least in the way we understand the term).

Not only are we the biggest organisation of our type in the world, but - or so we like to think - we are the best. We are certainly the only truly global cryptozoological research organisation, and we carry out our investigations using a strictly scientific set of guidelines. We are expanding all the time and looking to recruit new members to help us in our research into mysterious animals and strange creatures across the globe.

Why should you join us? Because, if you are genuinely interested in trying to solve the last great mysteries of Mother Nature, there is nobody better than us with whom to do it.

Members get a four-issue subscription to our journal *Animals & Men*. Each issue contains nearly 100 pages packed with news, articles, letters, research papers, field reports, and even a gossip column! The magazine is Royal Octavo in format with a full colour cover. You also have access to one of the world's largest collections of resource material dealing with cryptozoology and allied disciplines, and people from the CFZ membership regularly take part in fieldwork and expeditions around the world.

The CFZ is managed by a three-man board of trustees, with a non-profit making trust registered with HM Government Stamp Office. The board of trustees is supported by a Permanent Directorate of full and part-time staff, and advised by a Consultancy Board of specialists - many of whom are world-renowned experts in their particular field. We have regional representatives across the UK, the USA, and many other parts of the world, and are affiliated with

You'll find that the people at the CFZ are friendly and approachable. We have a thriving forum on the website which is the hub of an ever-growing electronic community. You will soon find your feet. Many members of the CFZ Permanent Directorate started off as ordinary members, and now work full-time chasing monsters around the world.

Write to us, e-mail us, or telephone us. The list of future projects on the website is not exhaustive. If you have a good idea for an investigation, please tell us. We may well be able to help.

We are always looking for volunteers to join us. If you see a project that interests you, do not hesitate to get in touch with us. Under certain circumstances we can help provide funding for your trip. If you look on the future projects section of the website, you can see some of the projects that we have pencilled in for the next few years.

In 2003 and 2004 we sent three-man expeditions to Sumatra looking for Orang-Pendek - a semi-legendary bipedal ape. The same three went to Mongolia in 2005. All three members started off merely subscribers to the CFZ magazine. Next time it could be you!

We have no magic sources of income. All our funds come from donations, membership fees, and sales of our publications and merchandise. We are always looking for corporate sponsorship, and other sources of revenue. If you have any ideas for fund-raising please let us know. However, unlike other cryptozoological organisations in the past, we do not live in an intellectual ivory tower. We are not afraid to get our hands dirty, and furthermore we are not one of those organisations where the membership have to raise money so that a privileged few can go on expensive foreign trips. Our research teams, both in the UK and abroad, consist of a mixture of experienced and inexperienced personnel. We are truly a community, and work on the premise that the benefits of CFZ membership are open to all.

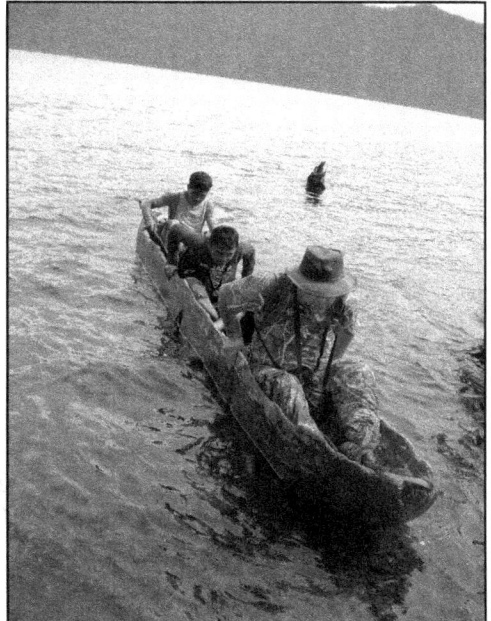

Reports of our investigations are published on our website as soon as they are available. Preliminary reports are posted within days of the project finishing.

Each year we publish a 200 page yearbook

We have a thriving YouTube channel, CFZtv, which has well over two hundred self-made documentaries, lecture appearances, and episodes of our monthly webTV show. We have a daily online magazine, which has over a million hits each year.

Each year since 2000 we have held our annual convention - the Weird Weekend. It is three days of lectures, workshops, and excursions. But most importantly it is a chance for members of the CFZ to meet each other, and to talk with the members of the permanent directorate in a relaxed and informal setting and preferably with a pint of beer in one hand. Since 2006 - the Weird Weekend has been bigger and better and held on the third weekend in August in the idyllic rural location of Woolsery in North Devon.

Since relocating to North Devon in 2005 we have become ever more closely involved with other community organisations, and we hope that this trend will continue. We have also worked closely with Police Forces across the UK as consultants for animal mutilation cases, and we intend to forge closer links with the coastguard and other community services. We want to work closely with those who regularly travel into the Bristol Channel, so that if the recent trend of exotic animal visitors to our coastal waters continues, we can be out there as soon as possible.

Apart from having been the only Fortean Zoological organisation in the world to have consistently published material on all aspects of the subject for over a decade, we have achieved the following concrete results:

- Disproved the myth relating to the headless so-called sea-serpent carcass of Durgan beach in Cornwall 1975
- Disproved the story

of the 1988 puma skull of Lustleigh Cleave

- Carried out the only in-depth research ever into the mythos of the Cornish Owlman.
- Made the first records of a tropical species of lamprey
- Made the first records of a luminous cave gnat larva in Thailand
- Discovered a possible new species of British mammal - the beech marten
- In 1994-6 carried out the first archival fortean zoological survey of Hong Kong
- In the year 2000, CFZ theories were confirmed when a new species of lizard was added to the British List
- Identified the monster of Martin Mere in Lancashire as a giant wels catfish
- Expanded the known range of Armitage's skink in the Gambia by 80%
- Obtained photographic evidence of the remains of Europe's largest known pike
- Carried out the first ever in-depth study of the ninki-nanka
- Carried out the first attempt to breed Puerto Rican cave snails in captivity
- Were the first European explorers to visit the `lost valley` in Sumatra
- Published the first ever evidence for a new tribe of pygmies in Guyana
- Published the first evidence for a new species of caiman in Guyana

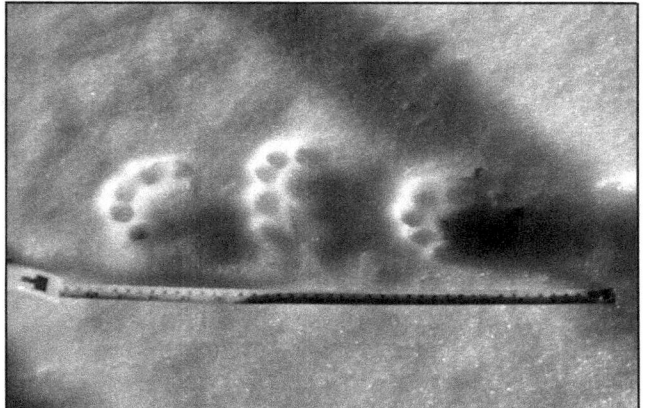

on a monster-haunted lake in Ireland for the first time
- Had a sighting of orang pendek in Sumatra in 2009
- Found leopard hair, subsequently identified by DNA analysis, from rural North Devon in 2010
- Brought back hairs which appear to be from an unknown primate in Sumatra
- Published some of the best evidence ever for the almasty in southern Russia

CFZ Expeditions and Investigations include:

- 1998 Puerto Rico, Florida, Mexico (Chupacabras)
- 1999 Nevada (Bigfoot)
- 2000 Thailand (Naga)
- 2002 Martin Mere (Giant catfish)
- 2002 Cleveland (Wallaby mutilation)

- 2003 Bolam Lake (BHM Reports)
- 2003 Sumatra (Orang Pendek)
- 2003 Texas (Bigfoot; giant snapping turtles)
- 2004 Sumatra (Orang Pendek; cigau, a sabre-toothed cat)
- 2004 Illinois (Black panthers; cicada swarm)
- 2004 Texas (Mystery blue dog)
- Loch Morar (Monster)
- 2004 Puerto Rico (Chupacabras; carnivorous cave snails)
- 2005 Belize (Affiliate expedition for hairy dwarfs)
- 2005 Loch Ness (Monster)
- 2005 Mongolia (Allghoi Khorkhoi aka Mongolian death worm)

- 2006 Gambia (Gambo - Gambian sea monster , Ninki Nanka and Armitage's skink
- 2006 Llangorse Lake (Giant pike, giant eels)
- 2006 Windermere (Giant eels)
- 2007 Coniston Water (Giant eels)
- 2007 Guyana (Giant anaconda, didi, water tiger)
- 2008 Russia (Almasty)
- 2009 Sumatra (Orang pendek)
- 2009 Republic of Ireland (Lake Monster)
- 2010 Texas (Blue Dogs)
- 2010 India (Mande Burung)
- 2011 Sumatra (Orang-pendek)

For details of current membership fees, current expeditions and investigations, and voluntary posts within the CFZ that need your help, please do not hesitate to contact us.

The Centre for Fortean Zoology,
Myrtle Cottage,
Woolfardisworthy,
Bideford, North Devon
EX39 5QR

Telephone 01237 431413
Fax+44 (0)7006-074-925
eMail info@cfz.org.uk

Websites:

www.cfz.org.uk
www.weirdweekend.org

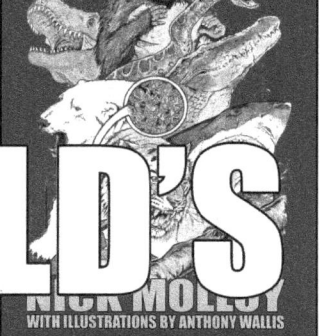

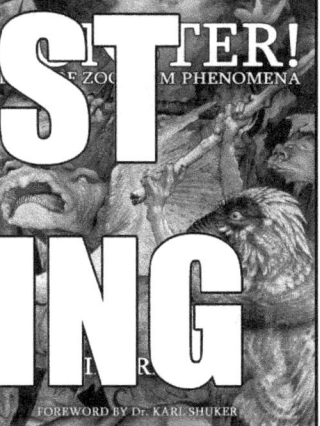
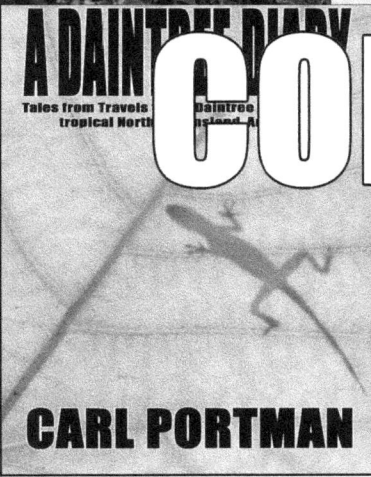
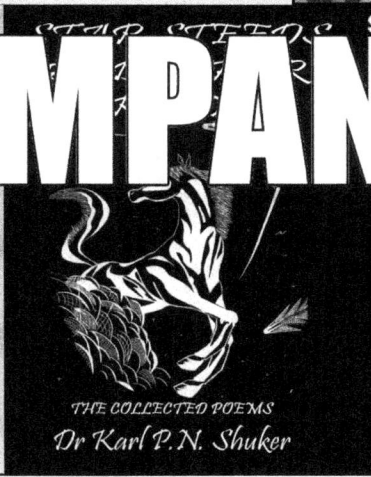

THE WORLD'S WEIRDEST PUBLISHING COMPANY

HOW TO START A PUBLISHING EMPIRE

Unlike most mainstream publishers, we have a non-commercial remit, and our mission statement claims that "we publish books because they deserve to be published, not because we think that we can make money out of them". Our motto is the Latin Tag *Pro bona causa facimus* (we do it for good reason), a slogan taken from a children's book *The Case of the Silver Egg* by the late Desmond Skirrow.

WIKIPEDIA: "The first book published was in 1988. *Take this Brother may it Serve you Well* was a guide to Beatles bootlegs by Jonathan Downes. It sold quite well, but was hampered by very poor production values, being photocopied, and held together by a plastic clip binder. In 1988 A5 clip binders were hard to get hold of, so the publishers took A4 binders and cut them in half with a hacksaw. It now reaches surprisingly high prices second hand.

The production quality improved slightly over the years, and after 1999 all the books produced were ringbound with laminated colour covers. In 2004, however, they signed an agreement with Lightning Source, and all books are now produced perfect bound, with full colour covers."

Until 2010 all our books, the majority of which are/were on the subject of mystery animals and allied disciplines, were published by `CFZ Press`, the publishing arm of the Centre for Fortean Zoology (CFZ), and we urged our readers and followers to draw a discreet veil over the books that we published that were completely off topic to the CFZ.

However, in 2010 we decided that enough was enough and launched a second imprint, `Fortean Words` which aims to cover a wide range of non animal-related esoteric subjects. Other imprints will be launched as and when we feel like it, however the basic ethos of the company remains the same: Our job is to publish books and magazines that we feel are worth publishing, whether or not they are going to sell. Money is, after all - as my dear old Mama once told me - a rather vulgar subject, and she would be rolling in her grave if she thought that her eldest son was somehow in `trade`.

Luckily, so far our tastes have turned out not to be that rarified after all, and we have sold far more books than anyone ever thought that we would, so there is a moral in there somewhere...

Jon Downes,
Woolsery, North Devon
July 2010

CFZ PRESS

Other Books in Print

Those Amazing Newfoundland Dogs by Jan Bondeson
Sea Serpent Carcasses - Scotland from the Stronsa Monster to Loch Ness by Glen Vaudrey
The CFZ Yearbook 2012 edited by Jonathan and Corinna Downes
Sea Serpent Carcasses - Scotland from the Stronsa Monster to Loch Ness by Glen Vaudrey
The CFZ Yearbook 2012 edited by Jonathan and Corinna Downes
ORANG PENDEK: Sumatra's Forgotten Ape by Richard Freeman
THE MYSTERY ANIMALS OF THE BRITISH ISLES: London by Neil Arnold
CFZ EXPEDITION REPORT: India 2010 by Richard Freeman *et al*
The Cryptid Creatures of Florida by Scott Marlow
Dead of Night by Lee Walker
The Mystery Animals of the British Isles: The Northern Isles by Glen Vaudrey
THE MYSTERY ANIMALS OF THE BRTISH ISLES: Gloucestershire and Worcestershire by Paul Williams
When Bigfoot Attacks by Michael Newton
Weird Waters – The Mystery Animals of Scandinavia: Lake and Sea Monsters by Lars Thomas
The Inhumanoids by Barton Nunnelly
Monstrum! A Wizard's Tale by Tony "Doc" Shiels
CFZ Yearbook 2011 edited by Jonathan Downes
Karl Shuker's Alien Zoo by Shuker, Dr Karl P.N
Tetrapod Zoology Book One by Naish, Dr Darren
The Mystery Animals of Ireland by Gary Cunningham and Ronan Coghlan
Monsters of Texas by Gerhard, Ken
The Great Yokai Encyclopaedia by Freeman, Richard
NEW HORIZONS: Animals & Men issues 16-20 Collected Editions Vol. 4 by Downes, Jonathan
A Daintree Diary -
Tales from Travels to the Daintree Rainforest in tropical north Queensland, Australia by Portman, Carl
Strangely Strange but Oddly Normal by Roberts, Andy
Centre for Fortean Zoology Yearbook 2010 by Downes, Jonathan
Predator Deathmatch by Molloy, Nick

The Owlman and Others by Jonathan Downes
The Blackdown Mystery by Downes, Jonathan
Big Cats in Britain Yearbook 2006 by Fraser, Mark (Ed)
Fragrant Harbours - Distant Rivers by Downes, John T
Only Fools and Goatsuckers by Downes, Jonathan
Monster of the Mere by Jonathan Downes
Dragons:More than a Myth by Freeman, Richard Alan
Granfer's Bible Stories by Downes, John Tweddell
Monster Hunter by Downes, Jonathan

TRADE MARK

BEWARE OF IMITATIONS

CFZ CLASSICS

CFZ Classics is a new venture for us. There are many seminal works that are either unavailable today, or not available with the production values which we would like to see. So, following the old adage that if you want to get something done do it yourself, this is exactly what we have done.

Desiderius Erasmus Roterodamus (b. October 18th 1466, d. July 2nd 1536) said: "When I have a little money, I buy books; and if I have any left, I buy food and clothes," and we are much the same. Only, we are in the lucky position of being able to share our books with the wider world. CFZ Classics is a conduit through which we cannot just re-issue titles which we feel still have much to offer the cryptozoological and Fortean research communities of the 21st Century, but we are adding footnotes, supplementary essays, and other material where we deem it appropriate.

Headhunters of The Amazon by Fritz W Up de Graff (1902)

Fortean Words

The Centre for Fortean Zoology has for several years led the field in Fortean publishing. CFZ Press is the only publishing company specialising in books on monsters and mystery animals. CFZ Press has published more books on this subject than any other company in history and has attracted such well known authors as Andy Roberts, Nick Redfern, Michael Newton, Dr Karl Shuker, Neil Arnold, Dr Darren Naish, Jon Downes, Ken Gerhard and Richard Freeman.

Now CFZ Press are launching a new imprint. Fortean Words is a new line of books dealing with Fortean subjects other than cryptozoology, which is - after all - the subject the CFZ are best known for. Fortean Words is being launched with a spectacular multi-volume series called *Haunted Skies* which covers British UFO sightings between 1940 and 2010. Former policeman John Hanson and his long-suffering partner Dawn Holloway have compiled a peerless library of sighting reports, many that have not been made public before.

Other books include a look at the Berwyn Mountains UFO case by renowned Fortean Andy Roberts and a series of forthcoming books by transatlantic researcher Nick Redfern. CFZ Press are dedicated to maintaining the fine quality of their works with Fortean Words. New authors tackling new subjects will always be encouraged, and we hope that our books will continue to be as ground-breaking and popular as ever.

Haunted Skies Volume One 1940-1959 by John Hanson and Dawn Holloway
Haunted Skies Volume Two 1960-1965 by John Hanson and Dawn Holloway
Haunted Skies Volume Three 1965-1967 by John Hanson and Dawn Holloway
Haunted Skies Volume Four 1968-1971 by John Hanson and Dawn Holloway
Haunted Skies Volume Five 1972-1974 by John Hanson and Dawn Holloway
Haunted Skies Volume Six 1975-1977 by John Hanson and Dawn Holloway
Grave Concerns by Kai Roberts

Police and the Paranormal by Andy Owens
Dead of Night by Lee Walker
Space Girl Dead on Spaghetti Junction - an anthology by Nick Redfern
I Fort the Lore - an anthology by Paul Screeton
UFO Down - the Berwyn Mountains UFO Crash by Andy Roberts
The Grail by Ronan Coghlan
UFO Warminster - Cradle of Contract by Kevin Goodman
Quest for the Hexham Heads by Paul Screeton

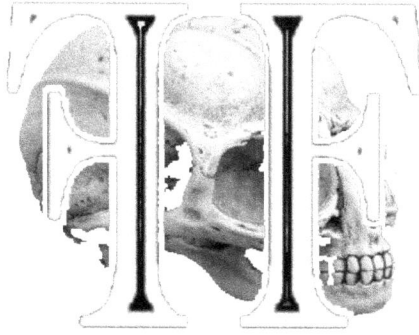

Fortean Fiction

J ust before Christmas 2011, we launched our third imprint, this time dedicated to - let's see if you guessed it from the title - fictional books with a Fortean or cryptozoological theme. We have published a few fictional books in the past, but now think that because of our rising reputation as publishers of quality Forteana, that a dedicated fiction imprint was the order of the day.

We launched with four titles:

Green Unpleasant Land by Richard Freeman
Left Behind by Harriet Wadham
Dark Ness by Tabitca Cope
Snap! by Steven Bredice
Death on Dartmoor by Di Francis
Dark Wear by Tabitca Cope

www.ingramcontent.com/pod-product-compliance
Lightning Source LLC
Chambersburg PA
CBHW072139270326
41931CB00010B/1817